GALAXIES

THE HARVARD BOOKS ON ASTRONOMY

Edited by HARLOW SHAPLEY and CECILIA PAYNE-GAPOSCHKIN

ATOMS, STARS, AND NEBULAE
Lawrence H. Aller

OUR SUN
Donald H. Menzel

EARTH, MOON, AND PLANETS
Fred L. Whipple

BETWEEN THE PLANETS
Fletcher G. Watson

STARS IN THE MAKING
Cecilia Payne-Gaposchkin

THE MILKY WAY
Bart J. Bok and Priscilla F. Bok

TOOLS OF THE ASTRONOMER
G. R. Miczaika and William M. Sinton

Harlow Shapley

GALAXIES

Revised by Paul W. Hodge

HARVARD UNIVERSITY PRESS

Cambridge, Massachusetts

1972

Preface

The 1943 edition of Harlow Shapley's *Galaxies* was the first book I ever bought on astronomy. Because of this and because it had such an importance to my education, the task of revising this book seemed at first a difficult and disrespectful act. Each page, even where outdated by more recent events, had a liveliness and personal quality that seemed to cry out against change.

I have tried to solve the problem by using Shapley's own example. When he revised *Galaxies* for its 1961 edition he found a happy middle ground between radical reworking and superficial updating. He kept the engaging flavor of the original, with its emphasis more on the giant steps in our understanding of galaxies than on details. He kept the unusual format of the original, with its very heavy tie to the nearby galaxies, and he kept the personal Harvard perspective, with its Cambridge view of the universe. I have tried to follow this pattern and still to keep the book well rounded and timely. More than one-third of the text is new material and 52 of the figures are new.

Besides the addition of new sections and chapters, the updating procedure has included a certain amount of revision of terminology. Shapley, as a pioneer explorer of galaxies, set many trends in the language of that branch of science. A few terms, however, have changed in their common usage since this book was first written, and I have in many places substituted the more modern choices for the original. For instance, where Shapley originally used the terms "cosmogony," "spheroidal galaxy," "Metagalaxy," and "cluster cepheid," I have substituted "cosmology," "elliptical galaxy," "universe," and "RR Lyrae variable," respectively. A good case can be made for Shapley's original choices of terms, but, unfortunately, scientists have not followed his lead in those few cases.

Throughout I have tried to keep the Shapley flavor, without, however, making what would have to be a futile attempt to imitate his writing style in the new sections and chapters. I hope that the compromise that results is readable and informative. Any success that this attempt has is due to Harlow Shapley. Any failure of style, any inaccuracy or omission should be accounted to me.

April 1971   P. W. H.

From the Preface
to the Revised Edition

HARLOW SHAPLEY

Since the writing of the first edition of this volume, which was based on the Harris Lectures at Northwestern University, there have been notable advances in the study of galaxies. Large new telescopes have gone into operation, in the Northern Hemisphere on Mount Hamilton and Mount Palomar, and in the south at Pretoria, Canberra, Cordoba, and the Boyden Observatory; the recording and analysis of galaxies is on the programs of all these instruments. The field is currently a very active one because it ties up locally with the Milky Way structural problems and distantly with the basic problems of cosmogony.

It would be possible to double the size of this volume by including a fuller report on investigations in current progress, and by going deeper into the technical details. I have treated only briefly, or not at all, such matters as the beginning of contributions from the radio telescopes, the nature and motions of the arms of spiral galaxies, the important extension to fainter stars of my early color-luminosity arrays for globular clusters, the evidence for the flattening and rotation of irregular galaxies such as that presented by Frank Kerr and G. de Vaucouleurs for the Magellanic Clouds, the details of the red-shift phenomenon, and the new views on the ages and evolution of galaxies and stars, derived from the new-era spectroscopic analyses of stellar atmospheres. These fields are being so busily developed at present that a statement of today may be outdated tomorrow. Most of what is here reported, however, will better stand the erosion of time.

Throughout this volume the estimates of distances and luminosities of galaxies have been substantially increased over the values presented in the first edition. Along with these changes have come new estimates of the rate of expansion in the Metagalaxy and of the elapsed time since the expansion began. These important revisions have stemmed from the recognition that our best distance indicators—the cepheid variables—are of at least two kinds: one suitable for the measures of the distances of globular clusters and cluster-type variables, and the other suitable for estimating the distances of the external galaxies and some of the classical cepheids in our own Galaxy.

The recent entry of radio astronomy and photon counters into the investigations of galaxies gives us a fresh view of our problem, a new look at the galaxies both as individuals and as units in the metagalactic structure. Many old metagalactic puzzles, such as the age of the universe, are being partially resolved through combining the techniques of radio and optical astronomy; but, inevitably, new problems, equally difficult, are emerging. Much work lies ahead for observer and theorist. This volume should be regarded as only an introductory step on one road to cosmic understanding.

1961

CONTENTS

Galaxies

1

Galactic Explorations

We who write and read these chapters are setting forth as explorers who rarely touch solid ground or come abreast of contemporary events. Scarcely anything as near as a naked-eye star or as recent as the discovery of America is to be considered. Most of the galactic radiation that comes to our eyes and photographic plates was generated thousands of years before man became curious about his universe. Nevertheless, ours is a practical exploration; the discoveries and deductions are of immediate concern to those who seek orientation in a complicated world. The stars and galaxies are linked with the sun, the sun with light, and light with our living and thinking on the earth's surface. Many times while on excursions into interstellar space we shall look back, objectively, at the planet where our telescopes are installed. But chiefly we shall be looking far outward into space, and remotely backward and forward in time.

In our swift excursions among the star clouds and galactic systems, we must pause frequently for detailed investigations. The tools of measurement and comprehension must be sharpened, and the new vistas studied in some detail to see where next our explorations can most profitably turn—to see, in fact, if we are getting anywhere. Investigations along the way should not be tedious. The side trips have an interest of their own; and there is a satisfaction in designing instruments for measuring stars and nebulae, as well as using them for measurement and interpretation.

Before we begin the exploration of the sidereal universe and the reporting of what is known about its more distant parts—an excursion that will lead from the Milky Way to the boundaries of measured space—it will be well to define galaxies, describe them preliminarily, and give an account of the way one measures such remote objects. We shall discuss single stars like our sun, describe and use groups of stars like the Pleiades and the rich globular cluster in Hercules, but devote most of our time and space to those yet greater star organizations—the other galaxies that lie beyond the bounds of the Milky Way.

Introductory

In the past as well as recently the great Andromeda spiral has played an important part in solving cosmic mysteries. It is the most conspicuous external galaxy except the Magellanic Clouds, and they are invisible to northern observers. It has been known from ancient times, and appears on the star charts of the Middle Ages. Its diffuseness, and the haziness of some of the brighter globular clusters, inspired Immanuel Kant two centuries ago to a speculation that is now justified. He suggested that at least some of these misty objects, scattered among the clean pointlike stars, might be other distant organizations, themselves composed of myriads of suns; they might well be considered "island universes" in the oceans of empty space, far beyond the confines of the Milky Way system in which our sun and all the naked-eye stars are embedded.

This island-universe hypothesis dimly persisted throughout the nineteenth century, notwithstanding skepticism on the part of a few important astronomers. Sir William Herschel, the founder, near the end of the eighteenth century, of the serious study of galaxies and star clusters, was hesitant. Others shared his doubts. Moreover, it was not an impelling theory; and although a great number of nebulous objects became known through the industrious telescopes of the Herschels, and after the year 1850 numerous spiral forms were discovered among these nebulae, little serious attention was given to the cosmic situation of the various nebulous types. Nobody worried much about them. Theories of the total universe remained dim and unprogressive.

A few astronomical writers of the nineteenth century (Nichol, Proctor, and von Humboldt, for example), playing imaginatively with the island-universe interpretation, introduced and used frequently the term "external galaxy" for those nebulous-appearing sidereal systems that seemed to lie outside our own flat Milky Way star-rich system. It was argued that if some of the faint nebulous objects are really stellar systems, hazy and unresolved because of distance, and if they are comparable in structure with our galactic system, they also could be and should be termed galactic systems, or galaxies. They should be differentiated from the diffuse nebulae, which are of gaseous nature and mostly lie among the stars of our own Galaxy. The Ring Nebula in Lyra (Fig. 1), the Orion Nebula (Fig. 2), the Crab—these are inherently nebulous. They are the true nebulae, and are clearly distinguishable, by location and composition, from the external galaxies. Indeed, many galaxies contain various sorts of these true nebulae as minor constituents, along with their stars and star clusters, dust clouds, and clouds of stars.

The cosmic stature and the location in space of galaxies was still uncertain in 1920. Even when their light was analyzed with that most revealing detector, the astronomical spectroscope, and the resulting spectra were found to be like that of the sun—an indication that they probably consist of stars and are not composed of diffuse gases, like the Orion Nebula—even then the astronomers of the nineteenth and early twentieth centuries remained uninterested, and unsure whether the spirals were inside or outside the galactic system. It was hard enough to locate in space the naked-eye stars and measure the nearest parts of the Milky Way. "Near things first" was the tacit and proper policy. Serious speculation on the horizons of the universe should wait.

Early in this century, however, came more precise and effective work in the measurement of stellar distances. The triangulation methods were much improved; then appeared the photometric methods that will presently be described. First, the photometry of eclipsing double stars began to extend our reach beyond the limits

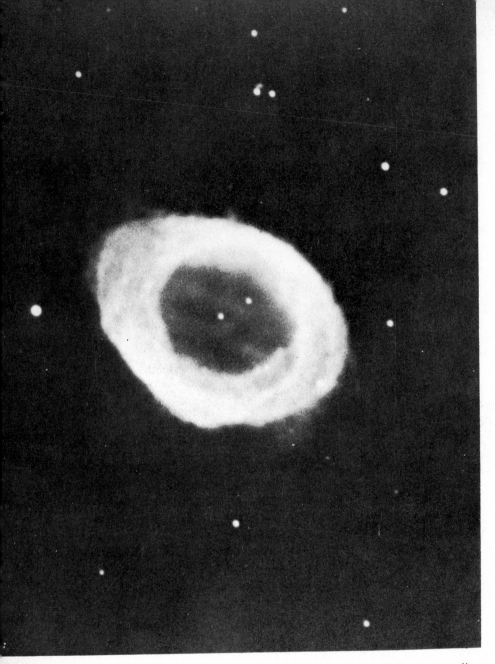

Fig. 1.   The Ring Nebula in Lyra. A true nebula, a gaseous ring surrounding a dying star. (Palomar photograph, 200-inch telescope.)

Fig. 2. The Orion Nebula photographed in red light, which gives a picture much like the drawings of a hundred years ago and quite unlike the blue-emulsion photographs now usually published. (Harvard photograph, Rocke-feller telescope.)

attained by triangulation; then came the estimation of stellar candlepowers, and stellar distances, by employing special characteristics in the spectra of stars; and soon thereafter was developed the powerful method based on cepheid variable stars to which we shall devote much of Chapter 3. With the advent of the measuring tool known as the period-luminosity relation for cepheid variables, the restriction of star-distance measurement to a few hundred light-years, that is, to a few million billion miles, vanished and we were ready to explore the Galaxy and the wide-open extragalactic spaces.

After various photometric criteria of distance had been developed, basically dependent on the cepheid variables, the globular star clusters were explored and their distances found to be astonishingly great. The concept of the galactic system as a stellar discoid with a greatest diameter of only a few thousand light-years was soon abandoned. Our Galaxy, or rather our idea of its size, grew suddenly and prodigiously. The clusters were shown to be affiliated with the galactic system, the center of which was found to lie some tens of thousands of light-years distant in the direction covered by the constellation Sagittarius. The heliocentric hypothesis of the stellar universe had to be abandoned, and the stage was set for further considerations of the nature and location of spiral galaxies.

Meanwhile the speeds of galaxies had been measured spectroscopically (Chapter 9), and the remarkable characteristic known as the *systematic* "red shift" was discovered. Its interpretation as an indicator of the general recession of the galaxies was widely but not unanimously accepted. For a time the evidence of a measurable angular rotation of the brighter spirals, shown by motions perpendicular to the line of sight and discernible by intercomparison of photographs separated by a short term of years, argued strongly against the acceptance of the great distances indicated by the interpretation of spirals as island universes quite outside our Milky Way system. At great extragalactic distances, any measurable cross motion would have corresponded to unreasonably and disruptively high actual speeds. But the measurements of angular rotation, or of cross motion of any kind, had been difficult and uncertain, and their evidence was discounted when individual cepheid variables were found by E. P. Hubble and others in the nearer spirals. These cepheids immediately made it possible, as we shall show later, to determine the stellar luminosities; and thereby the great distances were confirmed. Spiral nebulae, as H. D. Curtis and Knut Lundmark had ably argued, and as long before them Emanuel Swedenborg, Immanuel Kant, John Michell, and others had suggested, were finally proved to be galaxies, each with its millions of stars.

As soon as cepheid variables and other giant stars in the spirals had been identified, the same methods of distance measurement that some years earlier had been successfully used on the globular clusters were applied in detail by Hubble to the galaxies. The exploring of the space that lies beyond the boundaries of the Milky Way was actively begun. The periods and magnitudes of the cepheid variables, the apparent magnitudes of the brightest invariable stars, the total magnitudes of the galaxies, and their angular diameters—all now became criteria of the distances of galaxies, as previously for globular clusters. No new techniques were necessary; but large telescopes were required, and fast photographic plates, and more careful consideration of the standardization of methods through the more precise study of the motions and luminosities of nearby stars. An important revision of the first standardization is described in Chapter 3, and throughout this edition appropriate corrections to earlier values are applied. Later it transpired that the red shift (recession speed) is a rough indicator of distance.

So much for a rapid survey of a rapidly developing subject. A proper introduction to the galaxies should also contain a few further paragraphs on the techniques of measuring stellar distances and galactic dimensions. But first a digression to clarify the terminology. Fortunately, most of the language of the astronomy of galaxies is common talk. The explanation of technical words and phrases can therefore be brief.

Interruption for Definitions

Magnitude is the astronomical term for indicating brightness, not size. Actually, stellar magnitude is the logarithm (on a special base) of the faintness. The larger the magnitude numerically, the fainter the star. For example, a few of the brightest naked-eye stars are of the first magnitude; the faintest seen with the unaided eye are of the sixth magnitude; the faintest visible with the largest telescopes are near the eighteenth magnitude, and the faintest stars that such telescopes can now photograph, about the twenty-fourth magnitude. The difference between the *apparent magnitude* of a star (how bright it looks to us) and the *absolute magnitude* (how bright it would be at unit distance) becomes important in the measurement of distance, as is shown later in this chapter.

With a *spectroscope*, containing a prism or comparable light-dispersing device, the radiation from a star or galaxy can be spread into the various colors of which the light is composed, each color with its characteristic wavelength. (For an account of spectroscopes and other starlight-analyzing equipment, see another volume in this series, *Tools of the Astronomer*, by G. R. Miczaika and W. M. Sinton.) This streak or band of colored light—the spectrum—is generally crossed by dark lines produced by specialized light absorption in the stellar atmospheres above the principal radiating surface of the star. The pattern of absorption lines varies from star to star, and we therefore have many *spectral classes*, indicative of the various temperatures, sizes, densities, and atmospheric conditions. The most common spectral classes carry the designations *B, A, F, G, K, M*—a series arranged in the order of decreasing surface temperature. Subdivisions are indicated numerically, 0 to 9; thus: $A5$, $G0$, $K9$.

The difference between a blue photographic magnitude and a yellow photographic magnitude is the *color index*. Color indices and spectral classes are closely correlated. For example, large color indices indicate red stars of spectral classes K and M; the intermediate color indices, like that of the sun, are associated with stars of classes F and G; and the small color indices refer to bluish stars of classes B and A.

The lines in the spectra of many stars and galaxies are shifted away from their normal positions. Interpreted on the basis of the Doppler principle, this shift shows motion of the radiating object along the line of sight. The motion is away from the observer (*positive radial velocity*) if the spectral lines show a *red shift*—a shift, that is, toward the red end of the spectrum, toward the longer-wavelength side of the normal position of the lines; the object is approaching if the shift is toward the blue end of the spectrum (*negative radial velocity*).

Radial motions, along the line of sight, are commonly expressed in miles or kilometers per second. The cross motion, at right angles to the line of sight, called the *proper motion*, is measured in angular units—degrees, minutes, and seconds of arc—per year or per century.

Stellar distances are measured by *trigonometric* and *photometric* methods. The *parallax* of a star is the angle subtended at the star by the radius of the earth's orbit (about 93 million miles). The smaller the parallax, the greater the distance (Fig. 3). Represented in seconds of arc, the parallax is numerically the reciprocal of the distance expressed in the unit called a *parsec*. Thus a *parallax* of 1 *second* of arc (1″) corresponds to a distance of 1 parsec. A parallax of 0″.1 means a distance of 10 parsecs; 0″.001, 1000 parsecs, and

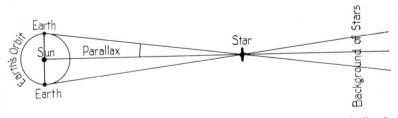

Fig. 3. A diagram that may help to define "parallax," the distance-indicating angle, and suggest the standard method of locating the nearer stars.

Fig. 4. The long-focus parallax-measuring refractor of the Van Vleck Observatory of Wesleyan University.

so on. A telescope devoted to parallax measurement is shown in Fig. 4.

Photometrically measured distances are those that are based on measures of the quantity of light. There are many kinds of *photometers*, some directly employing the human eye (visual), others used with the astronomical camera (photographic), and, most sensitive of all, those using phototubes (photoelectric) and photon counters. Ordinarily the photographic records of star brightness are made in blue light, but yellow-sensitive and red-sensitive photographic plates provide "yellow" and "red" magnitudes, and their use is increasing as the speeds of photographic emulsions are increased.

The distance light travels in a year, the *light-year*, is the distance unit most commonly used in this volume; it is equivalent to 5.88 trillion miles. One parsec equals 3.26 light-years, or a little over 19 trillion miles. The *kiloparsec* and the *megaparsec*, which are the convenient units for measuring the distances of galaxies, are a thousand and a million parsecs, respectively.

Right ascension and *declination* are the astronomer's usual cordinates for locating positions in the sky. They are analogous to longitude and latitude on the surface of the earth. Declinations north of the equator are positive; south, negative.

The *galactic longitude* and *galactic latitude* constitute an alternative system, which is used to indicate position in the sky with respect to the Milky Way. The equator of this system is the galactic circle; its north pole is in the constellation Coma Berenices, and its south pole is in Sculptor.

In the *numbering* and *naming* of star clusters, nebulae, and galaxies, the letters NGC and IC refer to the *New General Catalogue* of J. L. E. Dreyer and its indices, respectively (Chapter 8). A hundred of the brighter objects were previously catalogued by Chales Messier (Chapter 4), and bear his name (frequently abbreviated to M), as well as the NGC number. Thus the Andromeda Nebula is M 31 = NGC 224.

Cepheid variable stars, named for their prototype Delta Cephei, are now generally recognized as single high-luminosity stars that periodically swell up and shrink, with consequent periodic variations in light, color, temperature, spectral class, and other characteristics. The pulsations make cepheids easy to discover because their magnitudes are continuously changing. They are giant and supergiant stars and therefore stand out conspicuously among the brighter stars of a globular cluster or a galaxy. When the period of a cepheid is less than a day, the variable is commonly called a *cluster variable*, because stars of this sort were first found abundantly in globular star clusters (Fig. 5); a synonym is RR Lyrae variable.

Fig. 5.   The southern globular cluster No. 55 in Messier's catalogue. (Harvard photograph, Rockefeller telescope.)

The other two principal types of variable stars are the *eclipsing binaries* and the *long-period variables*, neither of which plays an important role in the discussion of galaxies.

The *novae* are relatively short-lived phenomena, but violent. They are stars that usually begin their variations explosively (Fig. 71); after a temporary brilliance they fade away gradually, but sometimes convulsively.

The *period* of a variable star is the average interval of time required for the variable to go through one complete cycle of its changes. The period is generally reckoned from maximum to maximum of light for cepheid and long-period variables, and from minimum to minimum for eclipsing binaries.

The *light curve* of a variable is a graph showing the course of the variations in brightness. Usually the stellar magnitude is plotted as vertical coordinate and time, in hours or days or period lengths, as horizontal coordinate. Typical light curves appear in Figs. 31, 43, 66, and 71. A *velocity curve* delineates the variations in velocity as revealed by spectrum-line shifts.

The Measurement of Distances to the Stars

Our isolation in the vacuum of surrounding space obviously prevents any use in stellar measurement of the surveyor's rods and chains; but the isolation does not hinder the use of the surveyor's triangulation method, at least for our planets and the nearby stars.

The terrestrial surveyor, when measuring the distance to some inaccessible mark, such as a distant mountain peak, establishes on the earth's surface a long triangle that has the mark at its vertex. He measures the length of the short side of the triangle and with telescopic pointings from the ends of this base line he gets the angles necessary for the solution of his terrestrial triangle and for the calculation of the distance to the selected mark.

For measuring the distance to a nearby star the sidereal surveyor sets up his base line in the solar system (see Fig. 3). He generally uses the diameter of the earth's orbit—a base line of 186 million miles; and his telescopic pointings at appropriate times of the year (usually at about 6-month intervals) give him the directions to the nearby star and provide the celestial triangle which is used to calculate the distance.

In principle, the triangulation for stellar distances is simple. In practice, it is very difficult; and it fails altogether, or is of little weight, when distances exceed 1000 light-years, chiefly because of instrumental limitations and inescapable observational errors. Indeed, to get reliable distances of 100 light-years requires exceedingly high accuracy of measurement. Under the inspiration and leadership of Kapteyn of Holland, and especially of Schlesinger of Yale, precise photographic methods of measuring angles on the sky have been developed, special telescopes have been constructed, and, since 1910, from the observatories in America, Europe, and South Africa have come relatively accurate trigonometric determinations of the distances of a few thousand stars. The work is basic. It is almost indispensable in preparing the tools for the measurement of the remoter parts of the universe; but such triangulation does not directly help us at all when we would survey the roads to star clouds and galaxies, for none of them is within its limited reach.

If the triangulation method alone were available, our task of measuring the distances to galaxies would appear hopeless. Looking at the Milky Way, we see, even with small telescopes, thousands of stars that are beyond our range; but how far beyond we would not know if trigonometry were our sole resource. Even some of the naked-eye stars are too remote for triangulation. *Indirectly*, however, the trigonometric measures of the nearby stars lead us to the Milky Way; they calibrate more potent methods.

It is fortunate that in the solar neighborhood the survey of the near and attainable stars can provide a straightforward standardization of the widely usable photometric method of measuring stellar distances. It might have been otherwise. There are many regions in the Milky Way where such standardization, or even the photometric method itself, would be complicated, precarious, and perhaps impossible. The chaotic dark and bright nebulae, such as that shown in Fig. 6, involve stars and planets that are badly situated as headquarters for reliable measurements of space.

The photometric method, like the trigonometric, is simple in principle but intricate in practice, and sometimes confusing and fallacious. In one sentence we can explain the method by saying that if in some way we know the real brightness of a star (its candlepower, or *absolute magnitude*), we can calculate the distance after we measure the *apparent magnitude*, which is a measure of the quantity of the star's spreading light that reaches us across intervening space. For example, $\log d = 0.2 \ (m - M) + 1$, where d is distance in parsecs, and m and M are apparent and absolute magnitudes, respectively. An uncertainly estimated δm must be added to the apparent magnitude to allow for dust-and-gas dimming. A familiar illustration is the estimating of the relative distances of street lights from their relative apparent brightnesses after we know that their candlepowers (actual brightnesses) are all the same. Dimness is an indicator of distance.

But there are difficulties. In the first place, the stars differ greatly in candlepower. Some are 10,000 times brighter than the sun, some 10,000 times fainter (Fig. 7), and not many of them are actually and clearly tagged with their absolute magnitudes (candlepowers). Also, the *apparent* magnitudes of stars are difficult to measure accurately; and the dust and gas of the space between us and the stars sometimes seriously dim the light in transmission, so that the simple formulas that apply to clear space do not hold.

Notwithstanding such difficulties, the photometric method is practicable and widely used; it can be applied not only to stars, but, as we have already noted, to globular clusters, and even to galaxies of stars. And once we learn to estimate accurately the total

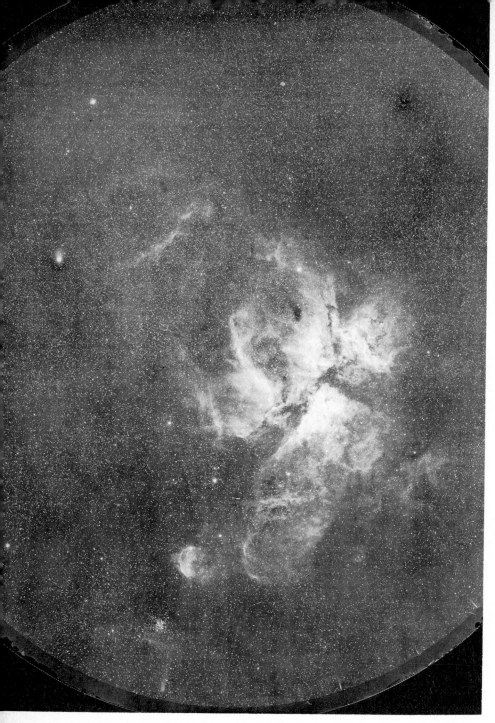

Fig. 6. Interstellar gas in the southern Milky Way near the star η Carinae. (Harvard photograph, ADH telescope.)

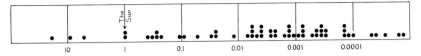

Fig. 7. Van de Kamp's plotting of the frequency of relative luminosities locates the sun with respect to the nearer stars; most of our neighbors are seen to be dwarfs in luminosity.

candlepower of an average galaxy—the total luminosity of its population of millions of stars—we can dispense with the use of individual stars such as cepheids, and, using instead this average candlepower of a whole galaxy, penetrate space with a new photometric measuring rod to distances a million times greater than those possible when our reliance was placed wholly on trigonometric methods. The principle of photometric measurement of stellar distances is as old as Newton, or older; but its practical development and use are recent and revolutionary. Without it, the existence of external galaxies would have remained hypothetical, and modern cosmology could not have existed.

The power of the photometric method lies in the developing of ways of estimating accurately the candlepowers of stars of those various types that are easily identifiable and are widely spread throughout the universe. To estimate great distances, we must first find the *absolute* brightnesses of stars that are of such great inherent radiance that they can be seen or photographed at such distances. Procedures relevant to this art will be described in Chapters 3 and 4.

As soon as we find the distances of the nearer galaxies, we can measure their angular dimensions and compute diameters in miles or in light-years. With sizes known, we proceed to intercompare the various galaxies. Finding much variety, we are immediately led to the setting up of types or classes. From that point the procedure is clear, and the questions ask themselves: What is the number of galaxies, how many kinds are there, what is their distribution in space, what are their relations to each other and to our own galaxy? Are they star factories? What of their internal activities, their forms, composition, origin, ages, destiny?

With such questions we are getting too far ahead of our story; moreover, the questions are too many, and at the present time too largely unanswerable. But in this preliminary chapter a description of the kinds of galaxies is appropriate, as is also a comment on their numbers.

The Kinds of Galaxies

Galaxies show much diversity in brightness and in size. There are titanic systems, such as the Andromeda Nebula, which are over a thousand times as bright and voluminous as dwarf systems like the Sculptor galaxy (Chapter 5). These two neighboring objects represent extremes in brightness, the average galaxy being about halfway between in absolute magnitude and number of stars.

Since there is a large spread in real brightness, there results in any survey of external systems brighter than a given apparent magnitude a preference for the most luminous. The dwarfs are systematically overlooked. We must be cautious, therefore, with such statements as "three-fourths of the galaxies are of the spiral type," because among the dwarf systems, most of which are not easily photographed in detail, the relative abundances of the various types may be otherwise than for galaxies of an average brightness, or for giant systems. Indeed, the *average* galaxy in a survey based on apparent brightness may differ considerably in type, as in luminosity, from the average galaxy in a given volume of space wherein everything is known.

To illustrate this rather important point by an astronomical analogy, we note that, when the stars visible to the unaided eye are listed in order of absolute luminosity, the sun appears to be less luminous than the average star. That position is certainly correct if we are dealing only with naked-eye stars—the sun is relatively dwarfish. But when we compare the sun with *all* stars, naked-eye and telescopic, that are in the neighborhood of the sun, it is found to be far above the average in brightness (Fig. 7). Careful research has revealed great numbers of dwarf stars. Probably the conditions found in our immediate neighborhood hold generally in the outer parts of the Milky Way, and possibly even in the distant central nucleus. Dwarf stars may predominate almost everywhere.

About 75 percent of the *brighter* galaxies so far satisfactorily observed and classified belong to the spiral type. The spirals generally show a bright nucleus, which is more or less spheroidal, and a flatter outer portion in which spiral arms are a conspicuous feature. The whole galaxy is watch-shaped, or more frequently wheel-shaped, with a conspicuous hub.

About 20 percent of the brighter galaxies are of the spheroidal or ellipsoidal type, radially symmetric about the center, or about an axis through the center, with indefinite boundaries, and no arms or other conspicuous structural detail.

The remainder of the brighter galaxies are irregular in structure and in form (the Magellanic type), or are peculiar variants on spiral and elliptical types.

The percentages 75, 20, and 5 for spirals S, ellipticals E, and irregulars I, respectively, are proper for a superficial survey of galaxies, such as for the thousand brightest as recorded in the Shapley-Ames catalogue. They are, however, not the correct percentages when a total count is made for a given volume of space, for then the fainter ellipticals and irregulars appear abundantly and the correct percentages may be more like 30, 60, and 10.

The illustrations that appear throughout this volume show that the spiral and elliptical galaxies can be readily subdivided. Knut Lundmark, of the Lund University Observatory in Sweden, and E. P. Hubble, of the Mount Wilson Observatory, among others, made classifications applicable to the brighter objects, and have introduced symbols and names to describe the various categories. These two systems of classification are similar in principle and detail; but the English astronomer J. H. Reynolds has appropriately emphasized the fact that practically every galaxy is distinguishable from all others; the classifications are only convenient shelves, not to be taken too seriously. (Classifications by Morgan and Mayall and by de Vaucouleurs are mentioned below.) The symbols used in this book for the brighter galaxies are those employed by Hubble but the names differ in some details; he used, for example, "extragalactic nebula" for our term "galaxy." Although by the writer and many others the term "galaxy" is preferred to "extragalactic nebula," the adjective "nebular" is sometimes found useful in referring to the faint nebulous objects that appear in the surveys of distant external galaxies. Strict and strained consistency of usage is avoided in the following pages.

Ellipticals. The brighter spheroidal or elliptical galaxies are subclassified at present according to the shape of the projected image. When classifying bright ellipticals, following Hubble's notation, we denote a circular image by $E0$; it may be the image of a truly spherical object, or of an oblate one viewed flat side on (Fig. 8). The oblateness of an elliptical galaxy is easily accounted for, if we accept a common hypothesis that flattening of a gaseous, liquid, or stellar spheroid is the result and measure of rotation about an axis. Gravity and centrifugal force combine to produce the distortion.

A very elongated elliptical system $(E7)$ is of course an extremely oblate wheel-shaped object seen on edge. (It is not at all likely to be a cigar-shaped object, because of the inherent instability of such a form under gravitational forces.) In the intermediate classes of elliptical galaxies, $E1$ to $E6$, tilt and oblateness are involved in varying degree. For any given object we can only say that it is at least as flat as it appears, and probably much flatter. Statistical arguments can be used to draw some rather uncertain conclusions about the relative frequency of the different degrees of flattening.

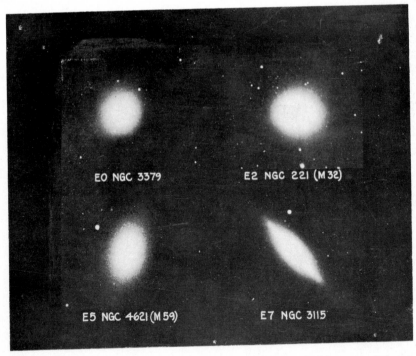

Fig. 8. Elliptical galaxies with four different degrees of oblateness. (Mount Wilson photographs by Hubble, 100-inch telescope.)

In the future, precise measures of the velocities within an elliptical galaxy, and a careful study of the degree of concentration of light along the different radii, may help to disentangle tilt from oblateness. At present, surmise does not seem important. The distribution of tilt can be more profitably studied among the spirals, which are unquestionably flat.

For distant galaxies it is difficult to discriminate between the elongated $E7$ elliptical system and some of the edgewise spirals that show little detail of internal structure and no granulation into stars, clusters, and nebulosity. In fact, when arranged in a sequence there is perhaps intrinsically little difference between these two types, and Hubble has proposed a connectant form, $S0$, representing the faintest discernible stage of spiraling.

Spirals. In classifying bright spirals, we start the series with Sa, in which arms can be clearly detected or reasonably surmised. There is little detail shown. The class Sb refers to objects with arms more distinct, perhaps more openly spread (Fig. 9). The structural detail and clustering, obvious in the arms of class Sb, become still more pronounced in the wide-open type, Sc, of which the great Messier 83 (Fig. 10), is a luminous example. We can go one stage further and use Sd for the most spread-out spirals (such as that shown in Fig. 11), in which the nucleus in conspicuous and the spiral arms are much broken and relatively dim. The symbols Sab, Sbc, and Scd denote intermediate forms. In all these spiral classes the arms emanate from the center of the galaxy, or from very near the center.

A few additional notes will complete this introduction to the galaxies:

1. There is a series of spiral forms, paralleling the normal type just described, in which the arms originate from the edge of a central disk rather than from a concentrated central nucleus. Frequently the arms start from the ends of a luminous bar that crosses the nucleus and surrounding disk (Fig. 12). These so-called barred spirals are designated SBa, SBb, and SBc, with $SBab$ and $SBbc$ for intermediates.

2. The *arms* are not the whole of the outer structure of the spirals; they are in fact merely enhancements on the background of light (stars) surrounding the nucleus, and are populated with young supergiant stars and nebulae. We shall see later that the total light contributed by what is generally recognized as the spiral arms is

Fig. 9. The terrestrial observer is almost exactly in the equatorial plane of this spiral galaxy, which bears the catalogue number NGC 4565. (Mount Wilson photograph, 200-inch telescope.)

Fig. 10. Number 83 in Messier's catalogue, a class *Sc* spiral, is convincing in its spectacular evidence that the universe is not static. (Harvard photograph, Rockefeller telescope.)

Fig. 11. Of the million galaxies recorded at the Harvard Observatory, this one is Number 4, discovered a century ago as a disappointment during a search for comets. The catalogue number is NGC 7793, the class is *Sd,* and the distance is about 5 million light-years. Outlying fragments of the galaxy are scattered over the whole field. (Harvard photograph, Bruce telescope.)

but a small portion of the total extranuclear light, much as coronal streamers seen at total solar eclipses are now known to be chiefly high lights on the background of the luminous globular corona.

3. Some external galaxies are without a visible central nucleus. Either no nucleus exists or it is hidden behind heavy obscuring dust.

4. All of the various subclasses, *Sa, SBc, E*5, and the others, include some forms considerably divergent from the average. The classification here used is obviously preliminary and useful chiefly for temporary guidance. Asymmetries exist; many intermediates lie between average forms, and chaotically irregular objects like the Magellanic Clouds are well known, as are also double galaxies (Fig. 13) and groups. An elaborate classification, based chiefly on form and arm structure, has been devised by G. de Vaucouleurs; and an important analytic classification involving spectrum features and

Fig. 12. NGC 7741 is a barred spiral with a complicated nucleus. (Palomar photograph, 200-inch telescope.)

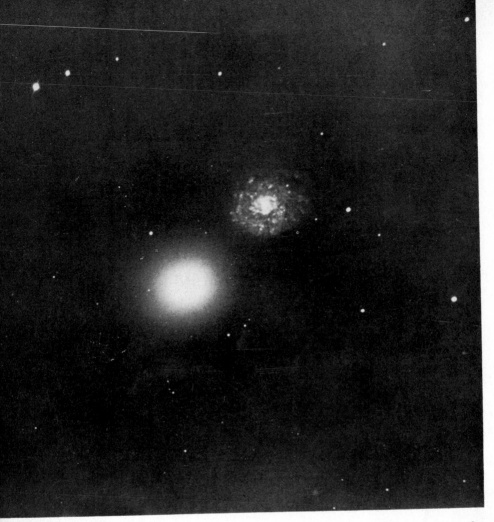

Fig. 13. Messier 60 and NGC 4647, a pair of unlike galaxies too remote for detailed analysis. (Mount Wilson photograph, 100-inch telescope.)

star types, as well as structure and degree of concentration, has been developed by W. W. Morgan, who bases his system largely on spectroscopic work by himself and N. U. Mayall and on Hubble's large collection of galaxy photographs.

5. The possibility of arranging the kinds of bright galaxies into one sequence, *E0–7, S0, Sa–d, I*, is not to be taken as a proof that one type develops into an adjacent form. We have here a "series of convenience," not an evolutionary tree.

It will appear later that hundreds of thousands of galaxies have now been photographed, but that only a few hundred are near enough to be studied in some detail. Perhaps a dozen or fifteen are within 2 million light-years (the census of dwarf galaxies is not complete, even to that small distance). In the vast material before us we find that the most satisfying of all galaxies for exploration are the Clouds of Magellan. The two following chapters will be devoted to these irregular systems. They are not only the nearest, but also the richest in procurable observational data, and in suggestion.

2

The Star Clouds of Magellan

The astronomy of galaxies would probably have been ahead by a generation, perhaps by 50 years, if Chance, or Fate, or whatever it is that fixes things as they are had put a typical spiral and a typical elliptical galaxy in the positions now occupied by the Large and Small Magellanic Clouds. If a spiral such as the one listed as NCG 4647 (Fig. 13) were, like them, only 160,000 light-years away, and its giant and supergiant stars were therefore easy to observe for motions and spectral characteristics, many of the dilemmas of the past 40 years would never have arisen. We should have known long ago whether spiral arms wind up, or unwind, or neither; whether they are superficial in galactic structure, or basic. And for more than 100 years we should have known that the spirals are star-filled external galaxies, and neither mysterious nebulous constituents of our own Milky Way system, as was once surmised, nor planetary systems in formation.

Similarly, if an elliptical galaxy like Messier 60 (Fig. 13) were only 160,000 light-years distant, we probably should have been spared many labors and doubts concerning such objects. Long ago we might have known, if we had given proper effort to the inquiry, something of the inner structure, perhaps even the laws of internal motions, in elliptical galaxies. We might have known definitely whether they are free of interstellar dust. We should have better understood their relation to globular clusters on the one hand and to the nuclei of spiral galaxies on the other. But, as it is, such problems are even now largely unsolved, because typical elliptical galaxies are very remote and therefore difficult to analyze.

The Andromedans have better luck! Those hypothetical investigators located in the great Andromeda Nebula (Messier 31) have two small elliptical galaxies close at hand, and the fine open spiral Messier 33 only half a million light-years away. They are spared our fortune of having as our nearest neighbors two irregular star clouds that for many years masqueraded before us as fragments detached from the Milky Way. Only gradually have we given these Clouds of Magellan the status of external systems and begun to appreciate that through their study we are analyzing an interesting but not very frequent form of galaxy (Fig. 14). The study of their irregular structures and motions gives little help in solving problems of the nature and operation of elliptical and spiral galaxies. But we must make the best of what we have, and it will soon appear that the best is indeed good. It's marvelous.

Cape Clouds they were called by the fifteenth-century Portuguese navigators, who picked them up in the southern sky as their ships approached the Cape of Good Hope. These unprecedented "Little Clouds" were, in fact, of some navigational use because they and the south pole of the heavens are at the three vertices of a nearly equilateral triangle; that is, they help locate the south pole. The oddity of them was described by Peter Martyr: "Coompasinge abowte the poynt thereof . . . certeyne shynynge whyte cloudes here and there amonge the starres, like unto theym whiche are seene in the tracte of heaven cauled Lactea via, that is the mylke whyte waye."

Corsali reports: "Manifestly twoo clowdes of reasonable bygnesse movynge abowt the place of the pole continually now rysynge and now faulynge, so keepynge theyr continuall course in circular movynge, with a starre ever in the myddest which is turned abowt with them abowte .xi. degrees frome the pole."

Fig. 14. The Small Magellanic Cloud, R.A. 0^h 50^m, Dec. $-73°$. To the right of this important and useful galaxy is the giant globular cluster 47 Tucanae. (Harvard photograph, ADH telescope.)

Variously designated by the navigators, the peculiar objects became indelibly associated in the literature of astronomy with the Great Circumnavigator. Only occasionally does an astronomer resort to the names Nebecula Major and Nubecula Minor. Magellan's associate and historian, Pigafetta, described the Clouds officially, during the course of that first round-the-world tour of 1518–1520, and thus made it appropriate to attach the explorer's name to these nearby galaxies that we ourselves now propose to explore.

The smaller of the two Clouds lies in the constellation of the Toucan. The Large Cloud is chiefly in Dorado, the Goldfish. Both Clouds spread beyond the boundaries of the constellations in which they chiefly lie. They illuminate a region of the sky romantically touched with exotic birds and beasts, if we judge by the constellation names. The water snake, the phoenix, the flying fish are there with the flamingo, the chameleon, the Indian fly, and the bird of paradise. All are near the south pole of the heavens, where to most of us the constellations are unfamiliar.

It would have been more convenient if the Magellanic Clouds were situated much farther north. Their cosmographic position has delayed their exploration, for there are ten observatories in the Northern Hemisphere to one in the Southern. Up to 1940 the southern stations of two American observatories, Harvard and Lick, had to do practically all the work on these important systems. The convenience of terrestrial astronomers obviously was not consulted in laying out the universe.

The Harvard-Peruvian Explorations

When the Clouds of Magellan are observed with the unaided eye, or visually inspected with any of the southern telescopes, or photographed with only moderate power, they appear not very large. The Small Cloud is then recorded as less than 4° in diameter; the Large, less than 8°, and dominated by a densely populated off-center bar or axis. Both are comparable in apparent size with some of the individual star clouds in our Milky Way, such as the bright patches of galactic light in Cygnus, Scutum, and Sagittarius. Such unpenetrating early views showed, nevertheless, a considerable amount of irregularity in form and in star density; but they were not especially revealing. Even as late as the beginning of this century the French writer Camille Flammarion summed up knowledge of the Large Cloud by saying that it contained 291 distinct nebulae, 46 clusters, and 582 stars. Descriptions such as this gave little suggestion of the deep significance and the tremendous richness of our nearest external galaxy, for they merely reported the occasional observations by Sir John Herschel and by a few other scientific voyagers to the Southern Hemisphere.

It was not until the growing Harvard Observatory had an opportunity in the eighteen-nineties to develop a southern station, and also had the good fortune of a substantial gift from Miss Catherine Bruce of New York, that the Magellanic Clouds began to unfold their story and inaugurate the astronomy of the galaxies. The Bruce photographic refractor (Figs. 15 and 16) came into existence through the cooperative efforts of Alvan Clark and Professor Pickering and their colleagues; it was in operation in Peru before 1900. This 24-inch large-field refractor, an exceedingly powerful instrument for its day, is now on the shelf. Its revised mounting carries

Fig. 15. The Bruce telescope building (and El Misti) at the Arequipa Station of the Harvard Observatory, 1899–1926.

the novel ADH Baker-Schmidt astrograph, one of the most effective of southern telescopes for the study of stars, star clouds, and galaxies.

There were many urgent jobs 70 years ago for the new Bruce refractor, which could photograph stars considerably fainter than the sixteenth magnitude in an hour's exposure, and could cover a field, on a single photographic plate, as extensive as the bowl of the Big Dipper. It had the responsibility of covering the whole southern sky—of doing much pioneering work along the rich southern Milky Way. For the Magellanic Clouds, therefore, the program proceeded slowly and it was several years before anything more significant was observed on the photographic plates than large numbers of star clusters and gaseous nebulae, such as were expected from the earlier visual observations by Sir John Herschel and others, and not hundreds but tens of thousands of stars.

The Clouds had been looked at for 400 years, but only now at the turn of the century were they beginning to be clearly seen. They were at last being accurately observed, but not by an ardent stargazer on the quarter-deck of an exploring frigate; not by the celes-

Fig. 16. The Bruce telescope in Cambridge before its going to the Southern Hemisphere.

tial explorer at his temporary observing station in Australia, South Africa, or South America; not even by the Harvard astronomer laboriously exposing large photographically sensitive plates in a powerful camera at the foot of El Misti in Peru. The Clouds were first really being seen by a young woman sitting at a desk in Cambridge, Massachusetts, holding in her hand an eyepiece with which she could examine a confusion of little black specks on a glass plate.

Miss Henrietta S. Leavitt of the staff of the Harvard Observatory had the gift of seeing things and of making useful records of her measures. She began by finding in the Magellanic Clouds the miracle variable stars that have subsequently turned out to be extremely significant both for the exploration of extragalactic space and for the measurement of star distances throughout our own Milky Way system.

She and other early workers on the Bruce plates had of course no way of knowing that the starlight from the Magellanic Clouds was some 1600 centuries old. In the first decades of the career of the Bruce telescope, distances of 160,000 light-years were quite unbelievable. But it is not uncommon for scientists to make systematic measures without knowing exactly what they measure. If the measures are good, those who make them can feel sure that significant interpretation will one day be forthcoming.

For Solon I. Bailey and Miss Leavitt, the two leaders in the discovery of the distant variable stars that were revealed by the Bruce telescope and the other Harvard Observatory instruments (Fig. 17), the immediate goal was the detection of variations in the intensity of starlight. Professor Bailey specialized on the star clusters, Miss Leavitt on the Clouds of Magellan. In 1906 she published a list of newly discovered faint variable stars in the two Magellanic Clouds—808 in the Large Cloud and 969 in the Small. The positions of these stars were recorded in appropriate coordinates, and

Fig. 17. The ADH Schmidt telescope in South Africa. The Bruce telescope at the Boyden station of the Harvard Observatory was replaced in 1952 by the Armagh-Dunsink-Harvard telescope, one of the most effective Schmidt telescopes in the Southern Hemisphere. (Photographed by the *Friend,* Bloemfontein.)

Fig. 18.   The bar of the Large Magellanic Cloud, R.A. 5^h 26^m, Dec. −69°. (Harvard photograph, 10-inch telescope.)

also their maximum and minimum magnitudes, referred to preliminary standards. It was not then noted that the range was generally about one magnitude, whether the variable star was among the brightest objects in the Clouds or among the faintest recorded by the photographic plate. There the matter rested for a bit, and we also shall let it rest until the next chapter, wherein come under discussion the astronomical tools that the astronomers have been able to fashion from the analysis of these nearby irregular galaxies.

Continuing the description of the Magellanic Clouds, we note that in addition to their very numerous variable stars there is within them a good sprinkling of many other stellar types. Their variable stars are duplicated in kind in our galactic system, even among the naked-eye neighbors of the sun. Similarly, the red giants of the solar neighborhood, the blue giants, and other highly luminous stars that have various spectral peculiarities, all find their counterparts in the Clouds. The Axis or Bar of the Large Cloud (Fig. 18) resembles somewhat the inner structure of some of the barred spirals.

Miss Annie J. Cannon's early work on the spectra of the brightest stars in the Clouds revealed among those of the common spectral classes a considerable number of important peculiar stars. In our own galactic system such stars are located chiefly, if not exclusively, in the thick of the Milky Way band—rarely, if ever, in the high latitudes, at large angular distances from the galactic circle. Since the Clouds stand well clear of the Milky Way, we can in consequence separate their peculiar stars from the abundant superposed ordinary stars of our own system. If a star of a peculiar type, such as a nova, a "classical" cepheid, a P Cygni or Class O star, is found in the direction of the Magellanic Clouds, we can say at once that it must be an actual member of the distant cloud and not a neighbor of ours—not a member of the intervening foreground of stars contributed by our own system. The same attribution to Cloud membership is possible for the loose star clusters and the gaseous nebulae found scattered over the Clouds. They do not belong to our own system because our loose clusters and gaseous nebulae are very rarely found so far from the Milky Way.

The Loop Nebula

The most conspicuous of the gaseous nebulae of the Magellanic Clouds, and in fact one of the two or three most spectacular objects of its kind known anywhere in the sidereal world, is the Loop Nebula in the Large Cloud, which bears the constellation designation 30 Doradus. We reproduce in Fig. 19 a photograph of this

Fig. 19. A detail of the preceding—the Great Loop Nebula as photographed at Bloemfontein with the Rockefeller telescope by J.S. Paraskevopoulos.

enormous gaseous structure—a picture made with the 60-inch reflector at Harvard's southern station at Bloemfontein, in South Africa.

The distance to the Large Magellanic Cloud is approximately 160,000 light-years (nearly a quintillion miles), and the linear diameter of the widely extended Loop Nebula is therefore astonishingly great. Let us compare it with the large nebula in Orion—a show object in our own Galaxy, about 1500 light-years distant. Both are visible to the unaided eye, the Orion Nebula appearing somewhat brighter. They have similar gaseous (bright-line) radiation; they are both associated with dense obscuring matter that conceals the stars lying beyond. They both have bright hot stars within them; and no doubt they owe to the high-temperature radiation of these included stars the energy that excites the gases to radiation. But the Orion Nebula is, in actual dimensions and in output of radiation, a pygmy compared with 30 Doradus. If the Loop Nebula were placed in the position of the Orion Nebula, it would fill the whole constellation of Orion, and the radiation from it and its involved supergiant stars would be strong enough to cast easily visible shadows on the earth. Radio telescopes have found a few like it in the distant, unseen reaches of our Galaxy, and far away in some other galaxies we have found comparable supergiant gaseous nebulae.

Not until some special photographs were made in red light, thus removing the emphasis on the blue radiations that are characteristic of bright nebulosity, did we discover, in the center of 30 Doradus, a cluster of 100 or more supergiant stars, spreading over an irregularly bounded volume, approximately 200 light-years in diameter. This gigantic cluster of giants is about 400 times as bright intrinsically as the great globular cluster in Hercules.

Magellanic Cloud Clusters

The Magellanic Clouds contain large numbers of star clusters. Three recent surveys of star clusters, carried out by Shapley and Lindsay at the Harvard and Armagh Observatories, by Lyngå and Westerlund at the Mount Stromlo Observatory, and by Hodge and Sexton at Berkeley, have together catalogued 1604 clusters in the Large Magellanic Cloud alone (Fig. 20). In addition, several hundred clusters have been recognized in the Small Cloud, many of them catalogued by Kron and Lindsay.

The question of the total number of clusters in the Magellanic Clouds is still very much an open one. There is no doubt that the majority of the Magellanic Cloud clusters have not yet been seen. There are four reasons that many clusters in the Clouds must have been missed by existing surveys. First, in the central core areas of the Clouds crowding of images is so great at faint magnitude limits that any cluster of faint stars there would be completely obscured by its background. Second, clusters that have stars primarily fainter than existing survey limits (approximately magnitudes fainter than 19) and that are not compact are missed because they do not show up on the plates as recognizable objects. The third reason for clusters being missed is that not all surveys to faint limits have covered the entire areas of the Clouds, which extend considerably beyond the obvious boundaries shown in most photographs. Finally, there is no doubt that interstellar extinction of light in the Magellanic Clouds occurs because of the presence there of dust, which is especially recognizable where it obscures portions of the bar of the Large Cloud, for example. This interstellar dust may completely obscure some clusters in the Clouds.

Is it possible to estimate the effects of these four limitations on surveys of clusters? Attempts certainly have been made. First, in order to determine the effect of crowding of images in the denser parts of the Clouds, it has been assumed that the distribution in size and age of clusters in the Clouds is not dependent on their

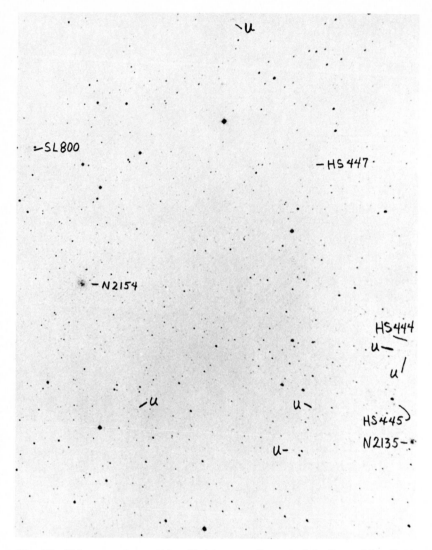

Fig. 20. Faint star clusters identified in a small area of the Large Magellanic Cloud. Numbers preceded by N are NGC numbers, numbers preceded by the letters SL are numbers assigned in the Shapley-Lindsay catalogue, numbers preceded by HS are from the Hodge-Sexton catalogue, and clusters marked U remain too faint to have been catalogued so far. (From Hodge and Sexton, *Astronomical Journal.*)

position, so that it is possible to gauge by comparison with less crowded areas how many clusters of different characteristics might be so obscured. For example, in the Large Magellanic Cloud it is found that a typical area of 160 square minutes of arc in the bar contains 24 clusters, 22 of which have stars brighter than visual magnitude 15.5. However, in an outer region of the Large Cloud an area of 5000 square minutes of arc near NGC 1866 contains 44 clusters, only 4 of which have stars brighter than magnitude 15.5. Thus is appears that about 90 percent of the clusters in the bar are missed because of crowding. This calculation indicates that a total of 2200 clusters might be missed because of the crowding of star images in the bar of the Large Cloud. If this figure is added to the total number of catalogued clusters with stars bright enough for them to have been catalogued by present surveys, the total number of clusters in the Large Cloud is found to be approximately 3800.

It is a more straightforward task to estimate the number of clusters missed because they have stars fainter than the faint limits of the surveys, about 19th magnitude. There are long-exposure plates available, taken with larger telescopes, which reach to fainter magnitudes. For example, an examination of a series of plates that have a limiting magnitude of 20.5 has shown that the catalogued clusters make up only about 60 percent of the true total number of clusters with stars brighter than absolute magnitude approximately +2. This leads to a revised total number of clusters for the Large Magellanic Cloud of about 5300, with stars brighter than absolute magnitude +2.

On the other hand, for the Large Cloud, at least, the search for clusters is now quite complete out to very large distances from the center of the Cloud and therefore only a very few clusters could have been missed because they lie beyond the boundaries of the searches. Perhaps 40 clusters may remain undiscovered for this reason.

Finally, it is much more difficult to estimate the number of clusters that are obscured by the interstellar dust in the Magellanic Clouds. To continue our example for the Large Cloud, studies of background galaxies indicate that there probably is a considerable amount of absorption of starlight, perhaps in a sheet of dust behind the more conspicuous star fields of the Cloud. However, information is presently too inconclusive to allow an estimate of the total number of clusters that might be obscured by this material. Perhaps a reasonable guess is that 600 or 700 faint clusters might be so obscured.

The conclusion for the Large Cloud is that the total number of star clusters must be on the order of 6000. This includes only clusters with stars brighter than absolute magnitude $+2$. If such faint clusters as NGC 188 or M 67 in our Galaxy also exist in the Large Cloud in large numbers, then the total number of clusters may be many more than 10,000. This is a remarkable number of clusters, especially when it is considered that in our Galaxy, a much larger, more massive system, there are only about 1400 catalogued clusters. Furthermore, in another irregular galaxy, IC 1613, which is near enough for star clusters to be detected and recognized, surveys have found not one cluster. Therefore, it appears that the Magellanic Clouds are exceptionally well endowed with clusters, for reasons that we do not understand. The number of clusters per unit volume in the Magellanic Clouds appears excessively large when compared with either our local system or certain other galaxies.

Most of the clusters in the Magellanic Clouds are of the open type, like the Pleiades or the Hyades. Nevertheless, there has been a very considerable amount of interest in those few clusters in the Clouds that appear to be globular clusters, like M 13 in Hercules or M 3 in Canes Venatici.

In 1930 the Harvard survey of clusters in the Magellanic Clouds listed only two globular clusters in the Small Cloud and only eight in the Large Cloud (Fig. 21). The brightest of all of these is the remarkable object NGC 1866, a huge cluster near the northern edge of the Large Cloud. In addition to those that seemed certainly to be globular clusters, there were 21 more that were tentatively identified as globular. However, evidence obtained at Harvard had cast doubts on the true identity of these objects as globular star clusters. For example, Miss Cannon's spectroscopic studies early in the 20th century showed that, whereas globular clusters in the Milky Way had spectra that resembled those of stars of late spectral type, spectra of clusters in the Magellanic Clouds gave unexpectedly early spectral types.

Further light was thrown on the subject of the globular clusters of the Magellanic Clouds in about 1950 when A. D. Thackeray took plates in two colors of the giant cluster NGC 1866 (Fig. 22). He discovered that the brightest stars, instead of being red as is the case in normal globular clusters, were very blue. At the same time Harvard plates were examined of this cluster and from them it was found by Shapley and Mrs. Nail that more than a dozen classical cepheid variable stars, all of about 3-day period, existed

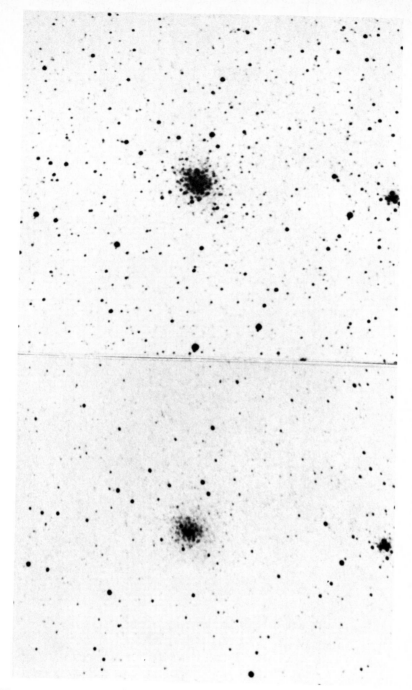

Fig. 21. Two photographs of the cluster NGC 1846 in the Large Magellanic Cloud. The upper photograph is taken in yellow light and shows the bright yellow giant stars, while the lower photograph is taken in blue light and shows that this cluster contains no bright blue stars. (Harvard photographs, ADH telescope.)

Fig. 22. The cluster NGC 1866, the brightest cluster in the Large Magellanic Cloud and a relatively young group of many thousands of stars. (Harvard photograph, Rockefeller telescope.)

in the cluster. No normal globular star cluster contains classical cepheid variables, because these objects are restricted to the flat spiral-arm portions of galaxies and are decidedly young stars. Thus the cluster, although it looks very much like a globular star cluster, contains the wrong kinds of stars. It appeared at that time to represent an entirely new type of object, completely unknown in our own Galaxy.

After the announcement of these discoveries, there was some question whether perhaps all of the globular star clusters in the Magellanic Clouds might be clusters of the new type. The Mount Stromlo Observatory telescopes were used in 1952 by Gascoigne and Kron to measure the integrated colors of these peculiar objects in an attempt to answer that question. Of those that they observed, roughly half were abnormally blue and the rest appeared normal. As a result of this apparent division, the two kinds of globular clusters in the Clouds were termed "blue globular" and "red globular" clusters.

Still, the colors measured in the early 1950s were not conclusive evidence that even the red globulars were normal clusters similar to those in our local system. Further evidence that this is so at least for a few came shortly thereafter when the two Radcliffe Observatory astronomers A. D. Thackeray and A. Wesselink found the first RR Lyrae variables (cluster-type variables) in the Magellanic Clouds. These were discovered in three of the red globular clusters, one of which belongs in each of the Clouds, with one lying somewhat between them in space. More recent studies of these clusters by Alexander at the Royal Greenwich Observatory and by Tifft using Mount Stromlo telescopes have shown that the RR Lyrae variables in these clusters are completely normal. Alexander's study of the Large Cloud cluster NGC 2257 turned up nearly 30 RR Lyrae variables and satisfied astronomers that this cluster is perfectly normal by the standards of those in our own Galaxy.

Color-magnitude diagrams, which display the properties of the cluster stars in a way that allows astronomers to identify the kinds of stars present and the cluster's age, for both the blue and the red globulars, became possible in the 1950s with the first use of very sensitive photoelectric photometers with the large Southern Hemisphere reflectors. Arp was one of the first to study cluster color-magnitude diagrams for the Magellanic Clouds, with very surprising results. He found for red clusters in the Small Cloud (he examined NGC 419 and NGC 361) that the stars were only superficially similar to those of normal globular star clusters. This led to the early suggestion that the clusters of the Cloud might be otherwise normal red globular star clusters, with the one difference that the abundance of metals was somewhat different from that of the typical clusters in our Galaxy. More recently Tifft found that the color-magnitude diagram for the faint star cluster NGC 121 did not show an anomalous character and appeared to be wholly consistent with the idea that this cluster is a normal globular star cluster. This is the one cluster in the Small Cloud that has known RR Lyrae variables and it may well turn out to be the only normal globular cluster in that galaxy.

In the case of the Large Cloud, color-magnitude diagrams have been obtained now for more than a dozen globular-like clusters, both red and blue (Fig. 23). Approximately 35 clusters have been

Fig. 23a. The intermediate-age cluster NGC 1831 in the Large Magellanic Cloud photographed by the Rockefeller reflector.

found to have color-magnitude diagrams that are roughly like those of normal globular clusters. These range from huge, very bright systems like NGC 1978 (Fig. 24) to tiny inconspicuous clusters too faint to have been included in the NGC catalogue. Studies by Sandage and Eggen, by Hodge, and more recently and most precisely by Gascoigne, have shown, however, that this similarity for most of the clusters is rather superficial. A few of the red globular clusters do appear to be normal, and these are the ones that are also found to have RR Lyrae variables, at least as far as we know. The others, of which NGC 1783 is a particularly well-

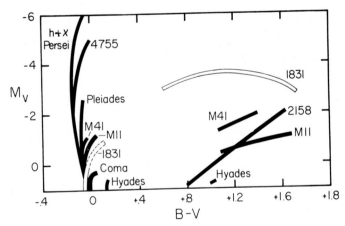

Fig. 23b. The color-magnitude diagram of the cluster NGC 1831 (open lines) compared with color-magnitude diagrams for clusters in our local Galaxy. (From Hodge, *Astrophysical Journal,* copyright University of Chicago Press.)

observed example, are quite abnormal, with the possibility that they are younger than normal globular star clusters, perhaps like the cluster NGC 7789 in our local Galaxy. These clusters may have ages of a few hundred million years, rather than the 10 billion years of normal globular star clusters.

Color-magnitude diagrams of the blue globular star clusters in both Clouds have been obtained by Arp, Gascoigne, Thackeray, and Hodge. All these clusters are found to have relatively normal unevolved stars but what appear to be rather abnormal distributions of red, evolved stars. It is quite clear that they are young objects and not in any way related to the old globular star clusters. They are unusual in the very large numbers of stars that they contain, in this respect being much more heavily populated than clusters of similar ages found in the solar neighborhood of our Galaxy. We do not know whether any such clusters do exist in our Galaxy. Possibly clusters of this type may lie in the distant regions of our Galaxy, perhaps closer to its center, where interstellar extinction completely obscures them from our sight. Perhaps we should take the point of view that, considering the extremely large number of clusters in the Magellanic Clouds, it is not surprising that a few will be abnormally rich and populous.

Fig. 24.   The Large Magellanic Cloud globular star cluster NGC 1978. (Harvard photograph, Rockefeller telescope.)

In any case, it is now believed that these blue globular clusters are in fact probably not different in any way from clusters of similar age in our Galaxy except in numbers of stars. For many years arguments raged over the possibility that what appeared to be peculiarities in the color-magnitude diagrams must be explained in terms of considerable differences in the mean chemical compositions of the stars involved. Certainly, it appeared most reasonable to attribute these apparent differences in color-magnitude diagrams to that cause, there being very little other possible explanation. More recently, detailed calculations of the evolutionary tracks of stars of clusters in this age range have thrown some doubt on the necessity of involving differences in chemical composition between our Galaxy and the Magellanic Clouds. There is a possibility that the differences merely arise from the superpopulation of the Magellanic Cloud clusters, which therefore give us a much more detailed picture of the later stages in the evolution of stars of that age than we get from the local clusters. Theoretical calculations of star models show that stars that enter these late phases go through several complicated gyrations, heating and cooling and expanding and contracting several times before finally dying as faint, small objects called "white dwarf stars." In order to see stars in all of these complicated stages of evolution it is necessary to have a cluster in which there are sufficiently large numbers of stars that a few are caught at each of these stages. Only in the Magellanic Clouds do we have clusters sufficiently populous for this to be true and it may well be that the peculiarities that we observe are the consequence of this richness. In this case it is the Magellanic Cloud clusters that give us the true picture (Fig. 25) and the local clusters with their insufficient numbers of stars would be the ones that give us the distorted view of the evolutionary stages of stars.

The Stars in the Clouds

The Clouds contain many varieties of stars, nebulae, and clusters; and although so near that even moderate-sized telescopes can show the individual objects, yet they are sufficiently remote that we may treat them objectively. In this lies their high value. With them it is possible, in a sense, to escape from the troublesome effect of differing distances—a serious obstacle in the intercomparison of stars in our own Galaxy.

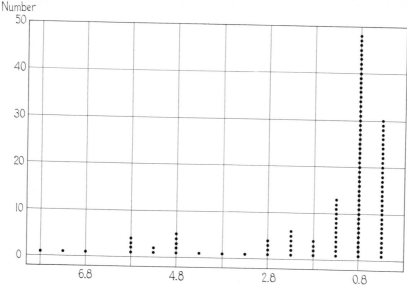

Fig. 25. Frequency of the angular diameters of the open clusters in the Large Cloud. One minute of arc corresponds, at the distance of the Cloud, to about 50 light-years.

We can safely adopt the principle that all the stars in a Magellanic Cloud are at approximately the same distance from the earth. Then, if we find that the Cloud stars of some specific type *appear* to differ in brightness among themselves, we know that it is because they really do differ. If the apparent magnitudes in the Clouds range, on the photographic plate, from 10 to 17, we can safely assume that the real luminosities of these stars also differ by seven magnitudes. It is otherwise in our galactic system.

To be sure, the Clouds have some thickness in the line of sight, and a star on the near side of the Cloud will be a bit brighter than if it were on the far edge; but the differences arising from location in the Cloud are small, relatively, and in our general analysis can be completely ignored. We can intercompare faint and bright stars in the Magellanic Clouds, knowing that we are intercomparing their candlepowers and not merely dealing with an illusion arising from conspicuously different distances, as is nearly always the circumstance when stars in our galactic system are intercompared.

Thus we may discover whether the peculiar class *O* stars in the Clouds are giants or supergiants in luminosity, or just average stars, because we can compare them directly with standard stars of known absolute luminosities, with the cepheids, for instance. The same procedure is possible for many other exotic objects that we should like to know about. The two Clouds turn out to be, therefore, a field in which to intercompare special or peculiar kinds of stars and nebulae. In pondering any particular evolutionary scheme, or development arrangement, we can now see quickly and certainly where a given object stands in the sequences of luminosities and masses; we can say responsibly whether it is perhaps a decadent star, or a primitive, or in mid-career.

We must as yet restrict our inquiries, however, to the candle-powers of giant and supergiant stars, because these nearby galaxies are so far away that no southern telescope yet constructed can photograph at their distances an ordinary star like our sun. We must deal chiefly with stars of from 10 to 10,000 times the luminosity of the sun, and solace ourselves with the reflection that most of the really interesting and active stars are giants of this sort. The sun and its kind are rather mediocre, and would be difficult to observe in detail from a distance greater than 50,000 light-years.

S Doradus and Other Supergiants

Before we leave the star clusters of the Large Magellanic Cloud, attention should be drawn to one irregular cluster that bears the catalogue number NGC 1910 (Fig. 26). It is nearly 400 light-years in diameter and contains a hundred or more giant and supergiant objects. One particular star, S Doradus, distinguishes this cluster, for it is possibly the most luminous star yet known in the whole universe, although to us, because of its distance of 160,000 light-years, it is considerably below naked-eye brightness. S Doradus is a variable star of an unusual sort, irregular in its light variations and of the peculiar P Cygni type of spectrum. With a light fluctuation from magnitude 8.2 to 9.4, its luminosity averages about 1,000,000 times that of the sun. It must certainly be a giant in size also, probably exceeding the orbit of the earth in diameter. Sirgay Gaposhkin has suggested that the star is actually double, with the equal components periodically eclipsing each other in a cycle of 40 years.

S Doradus is no doubt a blue, hot, highly efficient radiator (or pair of radiators). The many reddish supergiant stars distributed

Fig. 26. An area of the bar of the Large Magellanic Cloud showing the super-giant variable S Doradus, at the center of the circle in the loose clustering called NGC 1910. (Harvard photograph, Rockefeller telescope.)

throughout the Cloud give out nearly as much radiation, but, since they are relatively inefficient radiators, their diameters must in some instances be much greater—equal to that of the orbit of Jupiter—in order to provide sufficient radiative surface to maintain the enormous output of energy we observe. It is quite probable that in size many of these red giants in the Magellanic Clouds considerably exceed the greatest of the naked-eye stars near the sun, being bigger even than Antares and Betelgeuse. Certainly they much exceed those two stars in total radiation.

Distances and Dimensions

With the aid of the variable stars, as described in the next chapter, we have been able to determine the distance of the Magellanic Clouds as 160,000 light-years. There is some uncertainty in the allowance we have made for light-absorbing gas and dust in our own Milky Way and in the Clouds.

The Clouds are separated, center from center, by 21°, which corresponds roughly to 60,000 light-years. The distance from border to border is not more than the diameter of the Large Cloud. We can fairly propose that the two objects form a double system, faintly acting on each other gravitationally.

Actually the edges of the two Clouds are much closer together than they appear to be on first inspection. Special photographic plates, and a detailed counting of the faint stars, as well as diligent searching for outlying variables and open clusters, have greatly extended the recognized boundaries of both systems. In fact, each Cloud now appears to be a concentrated irregular mass of stars surrounded by a lightly populated envelope. That is, a haze of stars apparently surrounds the main body, which contains most of the mass of the system. Similar coronas of faint stars surround other galaxies, such as the irregular dwarf galaxy IC 1613 (see Chapter 5). The extent to which the boundaries of the Clouds have been pushed by the investigations carried on at the Harvard Observatory is indicated in Fig. 27. Evidence has been advanced by G. de Vaucouleurs, and by Frank Kerr and associates, using radio telescopes, to support the view that the Clouds are much flattened, and tilted from the line of sight by 30° for the Small Cloud and 65° for the Large—a suggestion that future radial-velocity work can test. The enveloping star haze is probably less flattened.

The extension in size of the Small Magellanic Cloud was first revealed on small-scale photographs of long exposure. The plates were made 50 years ago at an exploratory testing site in South Africa; some have exposures in excess of 20 hours. An extension, or wing, as indicated in Fig. 27, is directed toward the Large Cloud. It shows that the two systems may be nearly in contact by way of this faint stellar bridge.

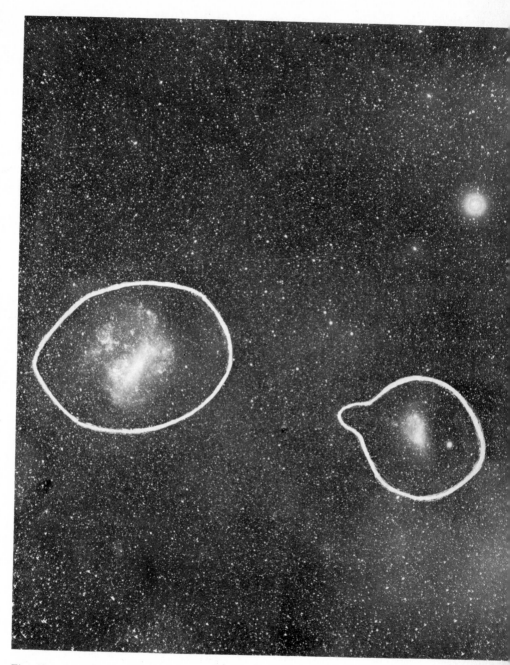

Fig. 27.   The overflow areas of the Magellanic Clouds, and the wing of the smaller. The bright star Achernar is also on the photograph. (Harvard photograph, AX camera.)

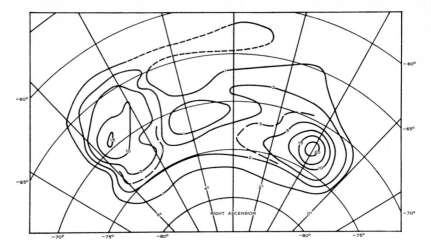

Fig. 28. A contour diagram showing the distribution of neutral hydrogen in the Magellanic Cloud system. The Small Cloud is at the right and the Large Cloud at the left. Orientation is similar to that of Fig. 27. (From Hindman, Kerr, and McGee.)

The recent work in Australia by the diligent radio astronomers there indicates that a hydrogen-gas envelope surrounds and permeates both Clouds (Fig. 28), so that a bridge of intergalactic gas connects them. It seems certain that both of these external galaxies lie near the edge of the star haze of our own galactic system. Their distances from the galactic plane are approximately 90,000 light-years for the Large Cloud and 110,000 for the Small. Some of our Galaxy's globular clusters and cluster-type cepheid variables are nearly as distant from the plane. One might consider the Clouds to be satellites of our much larger Galaxy; and certainly they are within its effective gravitational domain. These points are to be considered further in Chapter 5 when we examine the other neighbors of the Milky Way.

Motions of the Clouds

One is tempted to ask, without hope of an immediate answer, what has been the past career of these two ragged galaxies that are so near our dominating galactic system; what is to be their immediate future (in the next billion years), and their ultimate fate as units in the universe? Are they escaping from us, or coming in, or just tagging along? The partial answers now advanced help but little. No cross motions are as yet certainly detectable. W. J. Luyten, using Harvard photographs, showed that the cross motions must be exceedingly small.

The motions in the line of sight were first measured long ago by R. E. Wilson with spectrograms from the southern station of the Lick Observatory, and more recently at Pretoria and by the Australian radio astronomers. The Large Cloud recedes from the earth with a speed of about 170 miles per second, and the Small Cloud, which is farther from the galactic plane, with a speed of about 100 miles per second. But these figures represent chiefly our own rapid motion in our rotating Galaxy. They give the speed of our rotation about the nucleus in Sagittarius. Allowing for the rotational motion, we find that the line-of-sight speeds of the Large and Small Clouds in miles per second are about 0 and $+50$, respectively.

Evolution of the Magellanic Clouds

Although it is not possible at this time to determine where the Magellanic Clouds have been and where they are going, it is possible at least to trace back in history some of the evolutionary changes that have occurred in recent times. We cannot yet push this probing into their past very far, but studies have found powerful clues to the most recent periods in the histories of these objects. Using the cepheid variables as indicators of star-formation rates, C. H. Payne-Gaposchkin has shown that the two Clouds and the Galaxy have had separate and distinctive patterns of star formation over the last few tens of millions of years. The arguments are very complicated, requiring knowledge of the exact times that a giant star passes through an unstable phase, during which it becomes a cepheid variable, and the durations for each passage through this phase. It is also necessary to know how this changes for stars of different mass and how it shows up in stellar colors and luminosities. Through an extensive investigation of the immense collection of the Harvard plates of the Magellanic Clouds, Professor Payne-Gaposchkin was able to establish the luminosity curves and amplitudes for thousands of cepheid variables (Fig. 29). From these data she was then able to establish the rate of star formation at different periods in the recent histories of the galaxies.

Further light on the problem of the recent evolution of the

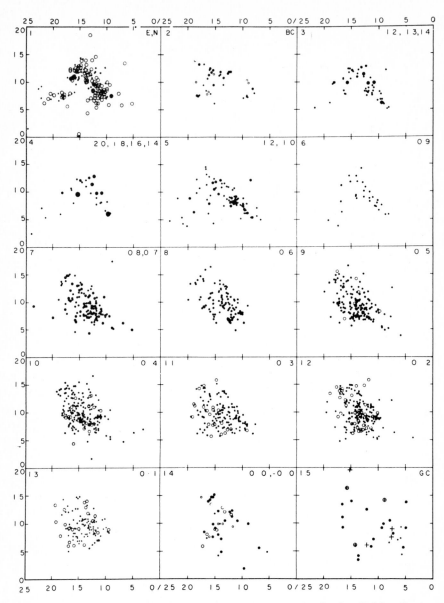

Fig. 29. The spatial distribution of different objects in the Small Magellanic Cloud: E, N refer to emission-line objects and blue supergiant stars; BC stands for blue clusters; GC stands for globular clusters. The remaining designations for the diagrams refer to cepheid variables of different periods, with the logarithm of the periods given in the upper right-hand corner of each diagram. (From Payne-Gaposchkin and Gaposhkin.)

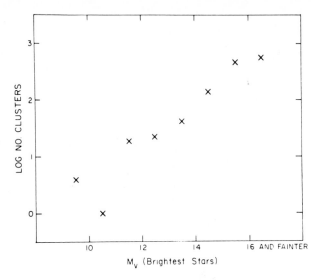

Fig. 30. The relation between the number of clusters in the Large Magellanic Cloud and the luminosities of their brightest stars. The left-hand side of the diagram is occupied by the few very young clusters and the right-hand side by the large number of old clusters. (From Hodge, Welch, Wills, and Wright, *Smithsonian Special Reports.*)

Magellanic Clouds was cast by an exploration of the pattern of cluster formation in the Large Magellanic Cloud carried out at Harvard and at the University of Washington. From measurements of the ages of 1200 of the Large Cloud clusters it was possible to establish the number of star clusters formed in different recent eras in that galaxy. The result was the discovery that star clusters in the Cloud appear to have been formed in bursts, lasting approximately 1 million years and covering an area of about 3000 light-years. These bursts seem to have occurred in the Large Cloud every 2 million years or so, appearing apparently randomly in different positions in the Cloud. These two lines of evidence indicate that the past histories of the Magellanic Clouds have not been continuous and uneventful but have included intense periods of activity, when large numbers of stars and star clusters were formed, with intervals in between of comparative inactivity (Fig. 30). In this further respect our studies of these remarkable little galaxies so close to our own have led us into knowledge and understanding that could never have come as easily from more distant galaxies.

3

The Astronomical Toolhouse

The two Clouds of Magellan, as remarked in the preceding chapter, are satisfactorily located in space for the effective study of many properties of galaxies, even though they are inconveniently far south for easy exploitation by the majority of astronomers. Their distance of about 160,000 light-years gives easy access to all of their giant and supergiant stars. Their considerable angular separations from the star clouds of the Milky Way keep them clear not only of most of the light-scattering dust near the galactic plane, but also of the confusingly rich foreground of stars and nebulosity near the Milky Way. They are nicely isolated.

During the past half century high profit has accrued from our studies of these nearby galaxies, for they have turned out to be veritable treasure chests of sidereal knowledge, and astronomical toolhouses of great merit. We shall see that the hypotheses, deductions, and techniques that arise from studies of the stars and nebulae of the Magellanic galaxies can be used to explore our own surrounding system, and also the more distant galaxies.

The usefulness of the Magellanic Clouds in the larger problems of cosmography can be illustrated by presenting, without stopping now to explain the meaning of the items or their significance, a partial list of the contributions to our knowledge of stars and galaxies that have already come from studies of the Clouds, or are on the way:

1. The period-luminosity relation;
2. The general luminosity curve, that is, the relative number of stars in successive intervals of intrinsic brightness;
3. Measures of the internal motions of irregular galaxies;
4. A comparison of the sizes, luminosities, and types of open star clusters;
5. The frequency of cepheid variation, shown by the number of cepheid variables compared with the numbers of other types of giant stars of approximately the same mass and brightness;
6. The spread of the lengths of period of cepheid variables;
7. The dependence of various characteristics of the light curves of cepheid variables on the length of period;
8. The dependence of a cepheid's period on location in a galaxy;
9. The total absolute magnitudes of globular star clusters, and the maximum luminosity of numerous special types of stars;
10. The demonstration of the "star haze" and "hydrogen haze" surrounding some if not all galaxies.

It seems inevitable that additional discoveries will reward the future investigators of these two external systems that can be studied objectively and in detail because of their nearness and externality.

Nearly all of the subjects listed can be investigated more successfully in the Magellanic Clouds than elsewhere. Any many can be read about in the technical reports better than here. Some involve the problems of stellar evolution; others, of galactic dimensions and structure. Several of the items will have their use chiefly in the future, rather than in the past; and although all are important in astronomy, only a few can be considered fully in this chapter.

The Abundance of Cepheid Variables

The outstanding phenomenon associated with the Magellanic Clouds is undoubtedly the relatively great number of giant variable stars, of which a majority are of the cepheid class (Fig. 31). They

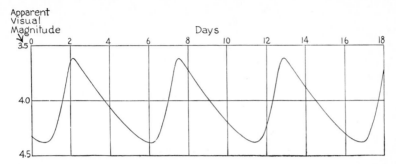

Fig. 31. An 18-day section of the light curve of the typical cepheid variable Delta Cephei, which for indefinite centuries will faithfully and monotonously repeat the 5.37-day oscillation.

are easily available for detailed investigation since they stand out conspicuously among the brighter stars.

In each of the Clouds there are more classical cepheid variables than are as yet known in our own much larger Galaxy. The survey in the Clouds approaches completeness; the survey in the galactic system is fragmentary and seriously hindered by the interstellar dust along the Milky Way where classical cepheids are concentrated. Probably fewer than half of the cepheid variables of our Milky Way system have been detected.

Of the variable stars in the Magellanic Clouds that have been worked up, about 80 percent are classical cepheids. In the neighborhood of the sun there are only a few of these pulsating stars; among them are Polaris and Delta Cephei, the latter being the star that gives a name to the class. In the solar neighborhood, as elsewhere in the galactic system, variables of other types are considerably more numerous than cepheids; for instance, here there are hundreds of eclipsing binaries, whereas only a few score are known in the Magellanic Clouds. Also we have found in our Galaxy more than 3000 "cluster" (RR Lyrae) variables, which are variables with periods less than a day, but only a few have been identified with certainty in the Magellanic Clouds. In the galactic system there are a great many long-period variables—the kind of stars that are carefully watched by the organized variable-star observers—but such stars are not yet abundant in the records of the Clouds.

Does this richness in the Magellanic Clouds of classical cepheid variables, with periods between 1 and 50 days, indicate that the population differs fundamentally in such irregular galaxies from that in the Milky Way spiral? Not necessarily so. The relative scarcity of cluster-type cepheids, long-period variables, and eclipsing stars in the present records of the Clouds is best accounted for by the relatively low candlepower of variable stars of those types. Even at maximum, such variables are not quite bright enough to get numerously into our eighteenth-magnitude picture of the Magellanic Clouds. Until recently we have photographed almost exclusively the giants that are 200 times or more brighter than the sun. The larger reflectors are beginning to explore among the fainter stars and possibly will soon reveal many cluster variables at the nineteenth magnitude, and eventually get down to stars of the sun's brightness.

The Period-Luminosity Relation and the Light Curves of Cepheids

Some years after Miss Leavitt had discovered and published 1777 variable stars in the two Clouds, she presented the results of a study of the periods of some of the variables. For the investigation she had selected the brightest of the variables as well as a few fainter ones. At once there appeared the interesting fact that, when the average brightness of a given variable is high, the period, which is the time interval separating successive maxima of brightness, is long compared with the intervals for fainter stars. The fainter the variable, the shorter the period.

The graph of her results for 25 variables is reproduced in Fig. 32. It is of historic significance. Miss Leavitt and Professor Pickering recognized at once that if the periods of variation depend on the brightness they must also be associated with other physical characteristics of the stars, such as mass and density and size. But apparently they did not foresee that this relation between brightness and period for cepheids in the Small Cloud would be the preliminary blueprint of one of astronomy's most potent tools for measuring the universe; nor did they, in fact, identify these variables of the Magellanic Cloud with the already well-known cepheid variables of the solar neighborhood. They simply had found a curiosity among the variables of the Small Magellanic Cloud.

Soon after Miss Leavitt's announcement of the period-magnitude relation for this small fraction of the variables that she had discovered in the Small Magellanic Cloud, Ejnar Hertzsprung and others pointed out that the nearby cepheid variable stars of the

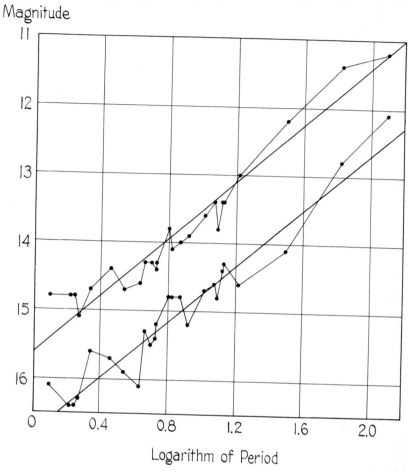

Fig. 32. Miss Leavitt's original diagram showing, separately for the maxima and the minima of 25 variable stars in the Small Cloud, the relation between photographic magnitude (vertical coordinate) and the logarithm of the period (horizontal coordinate).

Milky Way are giants—a fact that was readily deduced from their small cross motions and from spectral peculiarities. Therefore, if the galactic cepheids and the Magellanic variables are closely comparable in luminosities, these fifteenth- and sixteenth-magnitude objects in the Clouds must also be giants; and in order to appear so faint, they must be very remote, and so also must be the Clouds.

Shapley and others pursued the inquiry and supplemented Miss

Leavitt's work by studies of the variable stars that Bailey and others had detected in the globular star clusters. The many variables of the globular clusters are mostly cepheids of the cluster type with periods less than a day. But also in clusters are a few longer-period cepheids, and it was eventually possible to bring together all the data necessary for a practical but tentative period-luminosity curve. The new investigation appeared to connect the typical or "classical" cepheids with the cluster variables. Shapley then derived a zero point from trigonometric measures of the distances of the nearby cepheids and thereby changed the Leavitt relation from period and *apparent* magnitude to period and *absolute* luminosity, thus making distances determinable from light measures only, as will be shown below.

Using the apparently similar cepheid variables found in globular star clusters as guides to the zero point, Shapley derived the first period-luminosity relation for cepheid variables in 1917 (Fig. 33). He was able to establish a true and absolute luminosity for cepheids by taking advantage of the fact that what appeared to be normal cepheids existed in globular star clusters together with the cluster-type RR Lyrae variables. The latter are fairly common in our Galaxy and there are enough near the sun that the distances to the RR Lyrae variables could be established by statistical considerations based on their motions as seen over the sky. Therefore, it was well known that the RR Lyrae variables all have absolute magnitudes of approximately zero and thus Shapley was able to establish magnitudes for the cepheid variables in the globular clusters and, by comparison, also in Magellanic Clouds.

More recent studies of the true luminosity of RR Lyrae stars have confirmed rather well these early results utilized by Shapley more than 50 years ago. The accurate absolute luminosities obtained for globular star clusters for which we can measure distances by comparison of their main-sequence stars with main-sequence stars in nearby clusters, such as the Hyades, have led to the conclusion that in the mean, RR Lyrae variables have an absolute magnitude of approximately $+0.5$, with a spread from cluster to cluster of 0.2 or 0.3 magnitude. However, it is an entirely different story with regard to the longer-period cepheids found in the globular clusters. In the 1950s it was established that there are two kinds of cepheid variables, one belonging to the Population I, young, spiral-arm component of the Galaxy and the other belonging to

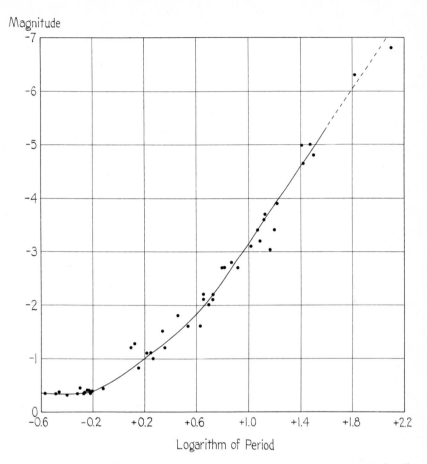

Fig. 33. The early Harvard period-luminosity relation, based on the 25 Small Cloud variables and cepheids from globular clusters in the local galaxy. The ordinate gives visual magnitudes on the absolute scale.

the Population II component, which includes the halo and the globular star clusters. The Population II cepheids were found on the average to be 1.5 magnitudes fainter for a given period than the Population I cepheids and therefore the period-luminosity relation for the two have a very different zero point. This meant that the cepheid variables in the Magellanic Clouds, which were clearly Population I cepheids, were very much more luminous than had originally been thought and therefore the distance to the Clouds must be twice as great as had been computed in 1917.

It may be well at this point to show how one uses the period-luminosity curve of Fig. 34 to measure the distances of the classical cepheids in our Milky Way, or the distance to some remote external galaxy, like the Andromeda Nebula. The procedure is very simple, once the period-luminosity relation is set up and accurately calibrated. First must come the discovery of a periodic variable star, and then, through the making of a hundred or so observations of the brightness at scattered times, comes the verification, from the shape of the mean light curve, that the variable belongs to the cepheid class. On a correct magnitude scale we next determine the amplitude (range) of variation, and the value of the magnitude half-way between maximum and minimum. The *median apparent magnitude, $\dot{m}$,* which is now almost always determined photographically or photoelectrically, constitutes one half of the needed observational material. The other half, namely the period P, is also determined from the observations of magnitudes.

With the period and its logarithm known, the relative absolute luminosity, $\dot{M}$, is then derived directly. For example, the simple formula

$$\dot{M} = -1.78 - 1.74 \log P$$

is satisfactory for getting the relative absolute magnitudes of all cepheids with periods between 1.2 and 40 days.

When we have thus derived the relative absolute magnitude from the period, we compute the distance d from the equally simple relation

$$\log d = 0.2 \ (\dot{m} - \dot{M} - \delta m) + 1,$$

where the distance is expressed in parsecs (1 parsec = 3.26 light-years, or about 19 trillion miles), and δm is the correction one must make to the observed median magnitude because of the scattering and absorption of starlight by the dust and gas of interstellar space. (The derivation of this standard formula is given by Bart J. Bok and Priscilla F. Bok in *The Milky Way,* and in various general textbooks.)

If space is essentially transparent, as in directions toward the poles of the Galaxy, δm can be set equal to zero. Such is the case for Messier 3 (Fig. 35). In directions where scattering is appreciable, we are frequently in trouble because δm is not zero and is difficult to determine. When we ignore the correction we have an upper limit for the distance. Thus, for cepheids in the Milky Way star

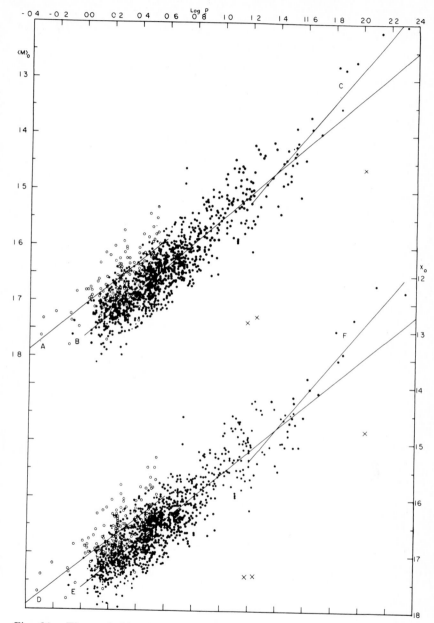

Fig. 34.   The period-luminosity relation for cepheid variables in the Small
Magellanic Cloud. The points are plotted twice, in the upper curve according
to luminosity at mean magnitude ($\langle M \rangle_0$) and in the lower curve according
to a mean magnitude (X_0), figured differently. Lines labeled *A, B, C, D, E, F*
are various solutions for the equation for the period-luminosity relation. (From
C. H. Payne-Gaposchkin and Sirgay Gaposhkin, *Smithsonian Contributions to
Astrophysics.*)

Fig. 35. The globular star cluster Messier 3, one of the most conspicuous in the northern sky, renowned for its nearly 200 RR Lyrae variables. (Palomar photograph, 200-inch telescope.)

clouds, where there is much dimming from dust, we can from this simple procedure, when scattering is ignored, determine only that the cepheids are not more remote than the computed distance; if δm is 1.5 magnitudes, they are actually only half as distant.

For a cepheid in a galaxy well away from the dust-filled Milky Way star clouds, we can safely assume that δm is less than 0.3, and rather accurately compute the distance of the cepheid from the formulas above. We then have not only the distance of the cepheid, whose M we get from P and whose $\dot{m}$ and P we get from the measures of magnitude, but also, without further measurement, we have the distance of the whole galaxy of a thousand million stars or more.

In summary, this simple but powerful photometric method based on cepheid variables involves only the observational determination of the periods and apparent magnitudes, followed by a direct calculation of absolute magnitudes and distances; we assume that our cepheid is of the classical type and not of the globular-cluster type.

Since cepheids with median photographic magnitudes as faint as the twenty-first can be discovered and studied with existing telescopes, and since such cepheids may have periods of 40 days, we can with the period-luminosity relation readily measure enormous distances. For example, a period of 40 days gives

$$\dot{M} = -1.78 - 1.74 \times 1.60 = -4.56$$

according to the first formula above. Then, away from the Milky Way absorption, the second formula gives

$$\log d = 0.2(21.0 + 4.56) + 1 = 6.112.$$

The distance that is measurable with this supergiant cepheid in high latitudes is therefore $d = 1,300,000$ parsecs, or approximately 4,200,000 light-years. The uncertainty in the result, on a percentage basis, is distinctly less than that in a measurement locally of 500 light-years by the older trigonometric method.

Periods of the cepheid variables of the Magellanic Clouds can be obtained by a comparison of measurements of brightness made over a sufficiently long time. In some recent studies made at Harvard it was possible to examine the behavior of cepheid variables in the Magellanic Clouds over a 70-year interval from 1896, when the first Harvard plates were taken, up to a very recent plate series.

Over a shorter interval it is very difficult to obtain an accurate measure of the periods of cepheids because they will have gone through too few observed cycles to allow a precise pinpointing of the times of each maximum or minimum. However, with the very long time available on the Harvard plate series it has been possible to determine periods of cepheids to many decimal places. For example, one of the recently measured cepheids in the Large Magellanic Cloud is HV 2369 (HV stands for Harvard Variable number); in a study of this star completed by Shapley and Mrs. Nail in 1952, it was possible to establish a period of 48.267 days. In a more recent study, carried out by Hodge and Wright in 1966, using a wider base line to determine the period, the period was more exactly determined at 48.28702 days. Another example is variable HV 5586, also in the Large Cloud. From a series of 30 plates taken in the years 1958 and 1959, it was only possible to establish that the period of this cepheid was 4.32 days. However, in 1966 Miss Wright completed an examination of its behavior over a 60-year interval on the Harvard plates and established its period more exactly at 4.3166344 days. Figures 34 and 37 show some recently obtained period-luminosity relations for cepheids in the Magellanic Clouds. The first shows the result of a wholesale survey of almost 2000 variable stars in the Small Magellanic Cloud carried out on Harvard plates by C. H. Payne-Gaposchkin and S. Gaposhkin at Harvard. The second shows the result of a much smaller-scale survey of a region in the Large Magellanic Cloud, in which somewhat more precise measurements were made on a series of plates of the region specially obtained for this specific purpose by Hodge.

There is not complete agreement from one survey to another on the characteristics of the period-luminosity relations of the two Clouds. Careful studies by Gascoigne at Mount Stromlo, using direct photoelectric means, generally confirm the zero point and the slope for the curve given in Fig. 24, but suggest that there are systematic errors in some of the recent photometry carried out in the Small Magellanic Cloud, and possibly also in the Large Cloud. The photometric problems of measuring precise magnitudes in the Magellanic Clouds have too often been underestimated. Because of the extreme crowding of images at the faint end and because of the presence of a diffuse unresolved background of stars, high accuracy in magnitudes is extremely difficult to obtain. Systematic

Fig. 36. A section of the Atlas of the Large Magellanic Cloud, showing identified cepheid variables with their Harvard Variable numbers. (Hodge and Wright.)

errors that can lead to misjudgment of the brightnesses of stars in the center of the plates as opposed to the outer parts can be as large as 0.5 magnitude. Any such error will give the wrong period-luminosity relation—either the wrong slope, or the wrong zero point, or both. It seems clear that we do not yet have well-established agreement on the characteristics of the period-luminosity relations within the Clouds, but the most recent careful photometric measures do agree fairly closely, well enough for us to feel confident in using the period-luminosity relation for distances that will at least be correct to within 20 percent.

To tie down the zero point, it is now generally accepted procedure to refer mainly to those cepheids in our local Galaxy that belong to galactic clusters. The distances to these galactic clusters can be measured precisely by comparing their main-sequence stars with comparable main-sequence stars in clusters that are near enough for us to obtain distances by geometric means (using parallaxes and the moving-cluster concept). Figure 37 shows recently derived

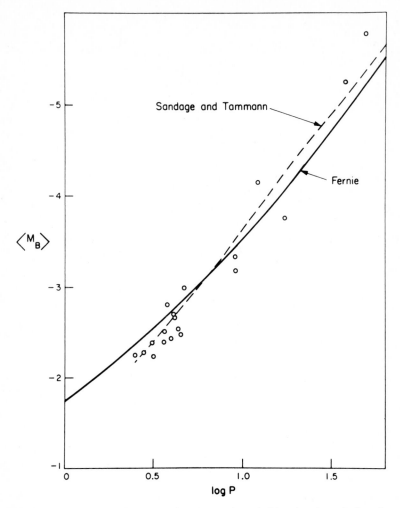

Fig. 37. A comparison between the observed period-luminosity relation for a small area of the Large Magellanic Cloud and the general period-luminosity relations derived by Fernie and by Sandage and Tammann. (From Hodge and Wright, *Astrophysical Journal*, copyright University of Chicago Press.)

period-luminosity relations that are based on these cluster members for the zero point and extragalactic cepheids for the slope ($\langle M_B \rangle$ in the curve is the mean blue absolute magnitude). One fit is a curved line, indicating that the usual formula for a period-luminosity relation that gives a constant slope over the entire period range cannot apply. The relations are compared in Fig. 37 with the periods and luminosities obtained in a recent Harvard survey of several Large Cloud cepheids and the agreement between the two seems to be very good, indicating that the Large Cloud cepheids in question fit the curves and that the absolute luminosities derived are probably quite valid and can give the distance to a reasonably high degree of accuracy.

There has remained over the years considerable concern with the possibility that we are making a grave mistake. Maybe the cepheid variables of the Magellanic Clouds are in reality somehow different from the cepheid variables in our local Galaxy and maybe therefore our conclusions about distances to the Magellanic Clouds and to other galaxies are based on a false identification. We have already seen how this happened years ago and are particularly concerned that we do not make the same mistake twice. It is clear that the cepheid variables have many features sufficiently alike in the different galaxies that they are at least very similar to each other in intrinsic properties. However, it is possible that small differences may exist, for example because of small differences in the mean chemical composition of the stars. These differences may cause a significant error in derived distances when they are ignored. Therefore, the slopes of the period-luminosity curves from one galaxy to the next have been examined with very great care. At first in the early 1960s, when several studies of the period-luminosity relations in various galaxies, using modern techniques, became available, it was thought that the slopes of the curves were in fact different from one galaxy to another. It looked as if in the Small Magellanic Cloud the period-luminosity curve was less steep that in the Large Cloud, and there was evidence that the slope was different in our Galaxy. Furthermore, some preliminary measurements of cepheids in another irregular galaxy in our local group of galaxies, IC 1613, suggested an even more gentle slope than in the Small Cloud. Great concern arose over the significance of this difference. Could it mean that the cepheid variables were in fact useless as distance indicators, or was it a difference that could in effect be corrected for if only we could discover the cause of it? If it was a difference due to metal abundance, would it be possible to carry out theoretical calculations for cepheid variables that could tell us just what the difference should be and why, thereby allowing us to avoid any difficulties? The answer seems to be that none of these is the case. Instead, it is believed that the difference in the slopes of the period-luminosity curves found in the early 1960s were entirely due to a combination of photometric errors and selection effects. If only a few cepheids are chosen in a study of a galaxy, it is possible to obtain an erroneous idea of the slope of the period-luminosity curve unless great care is taken in the selection of these cepheids. For example, if, as the temptation must often be, one chooses only the brightest stars, because they are easy to observe, then the period-luminosity relation will be biased. Systematically only the brightest stars for a given period will be involved and, for a relatively small sample of stars, this can lead to an invalid conclusion about the relation.

It is for this reason that several of the other properties of the cepheids have been scrutinized with care. If, in fact, the slopes of the period-luminosity curves from galaxy to galaxy are actually the same, as we now hope that they are, then is there any other difference that can be detected in cepheids from one galaxy to the next? Figure 38 shows another kind of relation that can be established by examination of the light curves of cepheid variables. This is the relation that exists between the period and the amplitude, which is the difference between the maximum luminosity and the minimum luminosity of the cepheid. It is found that in general the cepheids of longest period have the longest amplitude, whereas short-period cepheids can have very small amplitudes. In the Magellanic Clouds, where small-amplitude cepheids are difficult to detect and measure, we do not have a very broad sample of these objects; nevertheless, enough are known that some comparison can be made, both between the two Magellanic Clouds and between the Magellanic Clouds and our Galaxy, in this regard. Figure 38 illustrates the difference between the Large Cloud cepheids of a recent Harvard study and the conclusions obtained by Gascoigne and Kron on the basis of an exhaustive study of light curves of Galactic cepheids. There is a very conspicuous difference between the two galaxies over the entire period range. The Large Cloud cepheids of a given period tend to have larger amplitudes, sometimes by very considerable amounts, than do Galactic cepheids. Thus, we find differences, both between one Cloud and the other and between the Clouds and our Galaxy. Is this an intrinsic difference, due to fundamental differences in the properties of the stars in these different galaxies? Currently the most attractive answer to this dilemma is that proposed by Mrs. Payne-Gaposchkin, who believes that the cepheids are not intrinsically different, but that instead the differences in the period-amplitude relations that we

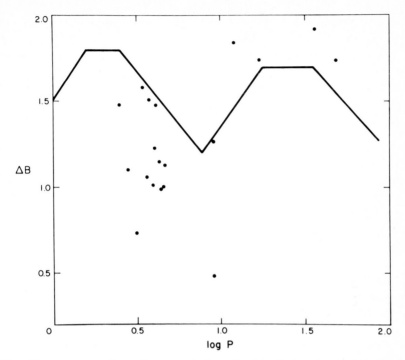

Fig. 38.   A comparison of the period-amplitude relation for several Large Magellanic Cloud cepheids (*dots*) with that for cepheids in our Galaxy (*line*). (From Hodge and Wright, *Astrophysical Journal,* copyright University of Chicago Press.)

find are the result of the different birth rates of cepheids at different times in the history of these three galaxies. She has shown, by some intricate calculations and comparisons, that what we observe is a possible and natural consequence of the fact that star-formation rates in the galaxies have been neither uniform in time nor the same from galaxy to galaxy. Because the amplitudes of cepheids depend very critically on their masses and because any cepheid of any given period can have one of a variety of different masses, and therefore ages, these factors all come together in a complicated way to mold our observed relations between amplitude and period. These relations then are an indication of the history of the forma-tion of stars that become cepheids in each galaxy, and, since this history is not necessarily dependent on the intrinsic properties (for example the chemical composition) of the stars, it is expected that we are safe in assuming that the period-amplitude relation differences do not argue against identifying the cepheids in the different galaxies as being the same objects.

Another way in which one can test the uniformity of cepheids from one galaxy to another is to examine their colors. In the early history of the study of the Magellanic Cloud cepheids, very little information was obtained on their colors because of the great difficulty in measuring accurately the colors of stars so faint. In the more recent studies, however, accurate determinations of the colors at various points in the light curves of the cepheids have become possible, and, furthermore, a comparison has been made between the mean colors of cepheids of different periods for the three galaxies. From the first reports on colors, there appeared to be a disturbing difference in the mean colors of the cepheids in the Magellanic Clouds from those in the Galaxy. It was reported that the Magellanic Cloud cepheids were anomalously blue, and that this difference could in no way be attributed to nonintrinsic differences such as higher amounts of reddening by interstellar dust in our Galaxy. More recently, however, it has been found that the

Fig. 39.   Four of the telescope buildings at the Cerro Tololo Inter-American Observatory in Chile, a modern center of Magellanic Cloud research. (Photograph by P. B. Lucke.)

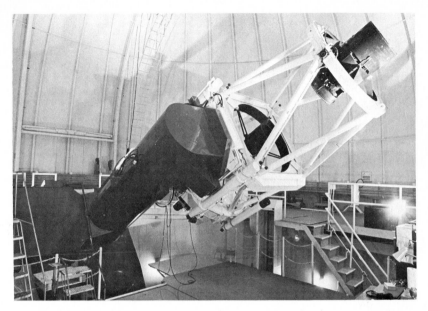

Fig. 40.   The 60-inch reflecting telescope at the Cerro Tololo Inter-American Observatory. (Photograph by P. B. Lucke.)

color data of the earlier studies may have been affected by system-atic errors both in the photometry and in the selection of cepheids. Gascoigne's 1969 study of colors in both Clouds confirms that the color differences first announced do not seem to exist, though there still seems to be a very small (about 0.1-magnitude) difference in the colors for the Small Magellanic Cloud cepheids, which appear to be just slightly too blue. This difference also may be erased in time, or conceivably it will be found to be the result of the different rate of cepheid formation in the Small Cloud as compared to the other galaxies.

What period is more frequent? This also is found to differ from galaxy to galaxy. In early studies of the Magellanic Cloud cepheids, Shapley and his many co-workers at Harvard were the first to point out that one of the most conspicuous differences between galaxies was the period-frequency diagram of the cepheids. They pointed out that for cepheids in our own Galaxy the most frequent period is slightly more than 4 days, whereas in the Small Magellanic Cloud the most frequent period is more nearly 2 days. Furthermore, it was found that the period-frequency diagram was different at different positions in the Clouds. For example, in the Small Cloud the most frequent period in the outer regions is under 2 days, whereas in the innermost regions, where the star density is very high, it is nearly 5 days. For many years this difference remained a mystery. Only when the calculations by theoretical means of the expected evolutionary paths of stars through the cepheid instability phases were refined was it possible to make sense out of these strange facts. It is now realized that cepheids of long period and high luminosity are stars of large mass, whereas cepheids of short periods and low luminosity are stars of relatively smaller mass. It is furthermore realized that stars of large mass, because they expend their energy at a very high rate, have relatively very short life expectancies as normal stars and rapidly evolve, becoming cepheids after only a relatively short time. On the other hand, stars of lower mass expend their energy at a very much lower rate (we can see this because their luminosity is low and luminosity is the way in which their energy is released) and therefore they can remain normal stars for a much longer time, not becoming cepheids until a much later date. Thus, cepheids with periods on the order of 50–100 days must be very young stars, only about 1 million years old, whereas cepheid variables with short periods, of only a few days, are older, with computed ages on the order of 100 million years. Therefore, the period-frequency diagram of a galaxy's cepheids is a picture of the

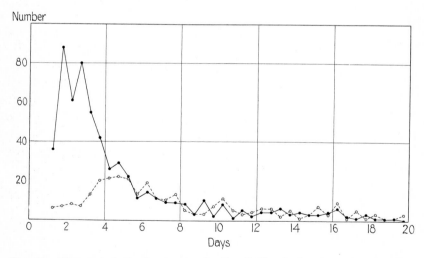

Fig. 41. The period distribution in the Small Cloud, showing great contrast with the distribution for the galactic system, which is represented by a broken line and open circles.

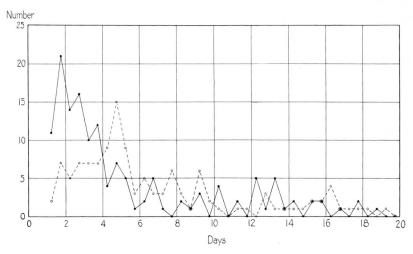

Fig. 42. Distribution of the periods of cepheids in the inner and outer regions of the Small Cloud. The broken line represents the former.

rate of formation of stars that become cepheids, and thus it is the most important clue regarding the variation over time of the formation rate of stars in a galaxy. We can see from Figs. 41 and 42, which compare period-frequency diagrams for the Small Cloud and the Galaxy, that because of the large number of cepheids with periods of about 4 days in our Galaxy there was apparently a burst of star formation corresponding to a date equal to the age of such cepheids, whereas in the Small Cloud a burst of star formation occurred earlier in the general field of that galaxy (corresponding to a date equal to the age of 2-day cepheids), though in the central parts of the Small Cloud there is a recent burst of star formation at an era determined by the ages of 5-day cepheids.

Finally, we can examine the shape of the light curves of the cepheids to see if they differ from one galaxy to the next. Many years ago, working with galactic cepheids, the Danish astronomer Ejnar Herzsprung produced evidence of a progressive change in the shape of the average light curve with period. The light curves for stars with periods around 10 days were found to be generally symmetric, while light curves for stars with periods of from 15 to 20 days showed greater amplitudes and were decidedly asymmetric. Figure 43 shows the way in which light curves vary in the mean for the Large Magellanic Cloud from short periods to long periods. These are typical light curves, though it is important to keep in

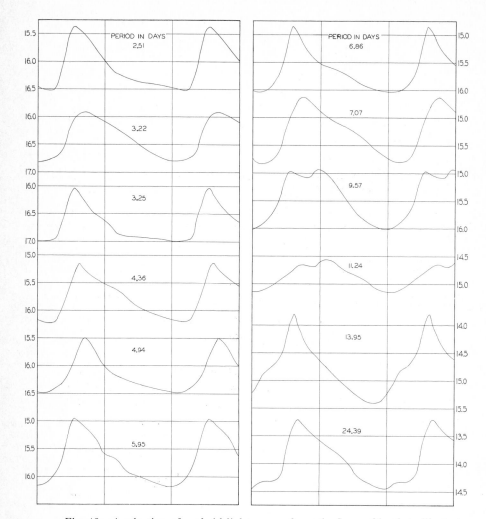

Fig. 43. A selection of cepheid light curves from the Large Cloud, to illustrate both the variety of curves and the peculiar form for the light curves when the periods are in the vicinity of 10 days. The horizontal scale, as usual, is in terms of time.

mind that for any one period light curves can vary over a fairly wide range of shapes. It is found that these pattern changes with period are followed in the case of all three galaxies, with only the very slightest difference. Most noticeably, it is found that the period for which double humps appear (shown in Fig. 43 for the cepheids of 9.57 days and 11.24 days) is different by approximately 1 day for the Magellanic Clouds and for the Galaxy. This is presently believed to be the only difference that cannot easily be explained in terms of nonfundamental properties of the stars, and it still remains a mystery whether or not this difference may be a clue to a fundamental difference, perhaps in chemical composition, between the Galaxy and the Magellanic Clouds.

In this discussion we have not mentioned one extremely difficult problem, one that gives considerable cause for concern in discussing Magellanic Cloud stars when we are after extremely precise luminosities and colors. This is the problem of interstellar reddening and extinction, caused by the general field of interstellar dust that pervades space in galaxies. We know that in the case of our Galaxy this dust occurs both in discrete obscuring clouds and in a sheet of roughly uniform density covering the entire disk of the Galaxy. Near the sun, the dust is sufficiently thick to cause a 0.75-magnitude dimming of starlight for stars that are about 3000 light-years away, and as much as 10 or more magnitudes for stars that are near the center of the Galaxy. We also know that this obscuration by dust causes an apparent reddening of starlight for distant stars, so that the measured colors have little relation to the true colors and temperatures of those stars. Toward the center of the Galaxy stars are found that are extremely red; they would be believed to be very cool stars if it were not for the fact that this reddening is due to the intervening dust and that the star spectra show clearly that they are in reality very hot stars. How much reddening and extinction by dust occurs in the Magellanic Clouds is of great concern if we are to use their cepheid variables and other distance indicators for purposes of establishing the distance scale in the extragalactic universe.

A great deal of argument and disagreement has focused on the question of the reddening and extinction of the Magellanic Clouds. It is shown clearly from surveys of background galaxies, very distant and small objects in all cases, that there is certainly dust present in the Clouds and that it must have some effect on the measured properties of Cloud stars (Fig. 45). Nevertheless, it is not yet established how much effect this dust has. From studies of many bright stars, primarily stars of spectral types O and B, it appears likely that only a small amount of obscuration and reddening is involved for the majority of the bright stars that we see, but we still do not know whether in fact there are many other stars in the Magellanic Clouds of this type, partly hidden behind an obscuring sheet of

Fig. 44. *Above:* Boyden Observatory on Harvard Kopje near Bloemfontein, Orange Free State, South Africa, where several photographic telescopes observe the southern stars and galaxies. *Below:* Another view, showing the 60-inch Rockefeller reflector.

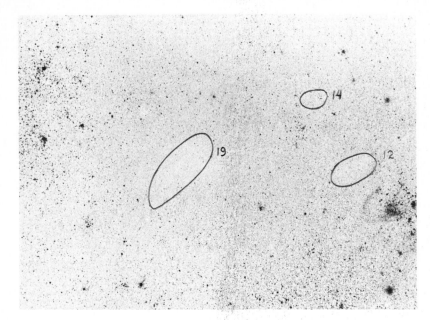

Fig. 45.   Three dust clouds in the Large Magellanic Cloud. Approximate boundaries of the dark nebulae are indicated.

dust, that we have not yet measured because of the obscuration. A present, it is found that only about 0.1 magnitude of obscuration seems to be affecting the colors and magnitudes of most of the O and B stars and the cepheid variables. This is not enough to cause great concern with regard to the period-luminosity relation because we can easily correct for such a small effect. However, the mystery still remains regarding the over-all total reddening and extinction due to the dust in the Magellanic Clouds.

The Luminosity Curve for Supergiants

After a census is made of a community of people or of stars, it is often of interest to know how the sizes or weights or brightnesses are distributed—how many individuals there are, for instance, of the various recorded diameters. A graph shows the distribution most plainly. A graph of the numbers of stars of a given type in successive intervals of absolute magnitude, plotted against the absolute magnitudes, is commonly called the *luminosity curve* for such stars. The luminosity curve may have one form for cepheid variables (featuring few supergiants, a great many giants, and no average or dwarf representatives), and a quite different shape for the red giant stars of the class of Antares and Betelgeuse.

The *general luminosity curve* lumps all types together. It is simply the frequency curve of the absolute magnitudes of all stars in a stellar system for which one does not bother, or is unable, to discriminate among types. It has a limited usefulness in the general study of stellar evolution; but, since there is for a majority of stars a fundamental relation between the mass of a star and its present luminosity (absolute magnitude), the general luminosity curve of a stellar system does yield some significant information on the frequency of individual masses and on the total mass of a stellar organization.

One of the more stubborn problems in the astronomy of our own Milky Way has been the derivation and analysis of the general luminosity curve. Because of our immersion deep in our own stellar system, we have difficulty in formulating a complete view of the Galaxy. We are bothered by natural preferences for nearby stars, or for highly luminous objects that are impressive though remote.

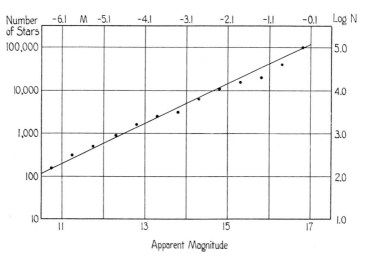

Fig. 46. The preliminary general luminosity curve for the brightest stars in the Large Cloud. Each point represents the total number of stars in the whole Cloud brighter than the corresponding magnitude. The vertical number scale is logarithmic. The straight line indicates, for example, that there are 10,000 stars brighter than photographic magnitude 14.6, which corresponds to absolute magnitude −4, approximately.

TABLE 1. *A preliminary census of supergiant and giant stars in the Large Magellanic Cloud.*

Absolute photographic magnitude	Total number of stars
Brighter than −6.5	735
−6.5 to −6.0	904
−6.0 to −5.5	1,460
−5.5 to −5.0	2,400
−5.0 to −4.5	3,260
−4.5 to −4.0	6,830
−4.0 to −3.5	11,080
−3.5 to −3.0	16,080
−3.0 to −2.5	23,170
−2.5 to −2.0	45,070
−2.0 to −1.5	103,350
All brighter than −1.5	214,370

And we are most bedeviled by the obnoxious gas and dust that foul interstellar space and mislead our measures of magnitude and distance.

The problem, however, can be attacked without these handicaps in an outside system like the Magellanic Clouds. We can there be sure, as we are not sure at home, that our survey of stars down to a given magnitude is complete. But the surety holds only for the supergiant and giant stars, because we are not yet able to reach effectively to the fainter objects, for example, to stars of the sun's brightness. In the Magellanic Clouds our luminosity curves extend, as shown in Fig. 46 for the Large Cloud, from the brightest super-giants, more than 100,000 times the solar brightness, to stars about 100 times as bright as the sun. In the near future the surveys will go much fainter, at least in special regions in the Clouds.

The census of the highly luminous stars has been made by counting in sample areas scattered throughout the Cloud. The areas are selected so as to give a fair representation of the very unevenly disposed population. The results are indicative rather than accurate; but they permit the construction of Table 1, which illustrates the great richness of this neighboring galaxy. Absolute magnitude −6.5 corresponds to apparent magnitude 12.1 in the Large Cloud and to a luminosity about 60,000 times that of the sun; absolute magnitude −1.5 corresponds to about 600 times the solar brightness. The fainter we go, the more stars there are. If the number in each magnitude interval were to increase with decreasing luminosity in the same way as in the sun's neighborhood, the population of the Large Cloud would rise to more than 40,000 million stars! This is an appalling number, but it is probably but a fifth of the number of stars in our own larger and more luminous galactic system.

The actual rate of increase in number of stars with decreasing brightness is, however, unpredictable. Except for the giant and supergiant stars represented in Table 1, the general luminosity curves for the Magellanic Clouds are undetermined and essentially unattainable. Our failure to reach easily the main-line average stars is one price we must pay for the advantage of being outside the Clouds.

Later it will appear that the Large Cloud is apparently a galaxy that is about average in dimensions and mass. But it does not follow that the luminosity curves, or the cepheid phenomena, are typical or average; they are quite different for elliptical galaxies and for the globular nuclei of spiral systems. Also the various characteristics of the star population vary from one part of a galaxy to another. Such heterogeneity prevails not only in distant spirals and in the Magellanic Clouds but also in our own Galaxy. The solar environment and the galactic nucleus are distinctly unlike. General luminosity curves (all star types together) for whole systems, therefore, do not amount to much in galactic interpretation, except perhaps for the elliptical galaxies and globular clusters where supergiant stars are rare and a smooth uniformity appears to prevail.

4

The Milky Way as a Galaxy

He who sees for the first time through a competent telescope the great star cluster in Hercules is definitely surprised, and naturally is skeptical when we offer the information, first, that each glittering point is a star far brighter than our sun, and second, that the whole concentrated globular assemblage of stars is so distant that the light now arriving has been en route for more than 300 centuries.

Star clusters of the globular form appeal to the imagination as well as the eye. They have enriched the general field of cosmography by contributing two important items to our knowledge of galaxies. They first indicated clearly that the sun and planets are eccentrically located in the Milky Way, far distant from the center in Sagittarius; and second, through their cepheids, they have helped to indicate the generality of the period-luminosity relation which first emerged as a tool for the measuring of sidereal distance in studies of another galaxy, the Small Magellanic Cloud.

Since the globular clusters have been useful in the portrayal of the Milky Way as a galaxy, we shall devote much of this chapter, which treats the Milky Way system as a cosmic unit, to discussions of clusters and of some strategically located variable stars.

Globular Clusters

The Hercules system has been so extensively studied during the past 60 years that we can now be sure of its great distance, its rich population, and the high luminosity of its brightest 10,000 stars. We know, for instance, that some of the stars are cepheid variables whose periods are correlated with absolute magnitude, and are therefore correlated with distance when apparent magnitudes are known. From the high background population of faint distant galaxies in the area around this cluster, we know that the intervening space must be quite transparent, and therefore that little or no correction to measures of magnitude and distance need be made on account of the absorption of light in space.

The Hercules cluster (Fig. 47) is commonly known by its number

Fig. 47. Number 13 in Messier's catalogue is the "Great Hercules Cluster," visible to the unaided eye. Its distance is about 30,000 light-years. (Harvard photograph, 16-inch telescope.)

in the famous catalogue of nebulae and clusters of various kinds, compiled around 1780 by the French comet hunter Charles Messier, who needed for his comet-seeking a list of those unmoving sidereal objects that are not comets but look hazily like comets and therefore in small telescopes are misleading. (Many modern amateurs, and some professional astronomers as well, have come upon one of these Messier objects and without consulting catalogues have excitedly telegraphed the supposed discovery of a new comet to the information bureau at the Harvard Observatory.) Messier did not care much about clusters and nebulae as such, and he catalogued a hundred of them as nuisances. He is remembered for this catalogue; forgotten as the applause-seeking discoverer of comets.

The individual stars in Messier 13 (the Hercules cluster) were, of course, not seen by Messier. "Nébuleuse sans étoiles," he records for the mighty Hercules swarm, and also for the 26 other globular clusters in his lists. It was left to the Herschels to resolve into stars most of the brighter globular clusters, and to the modern reflecting telescopes to resolve the faint ones.

The observers at the beginning of this century frequently suspected the existence of spiral arms and other structure in the brighter globular clusters, but these imagined structural details faded out of fact and memory with the increasing information given by the large reflectors, which show not hundreds of stars but tens of thousands. For a few clusters, wisps of dark nebulosity give a semblance of irregular form.

Practically all of the approximately 100 known globular clusters of our galactic system (Dr. Helen Hogg tabulated 106 in 1955) are smooth and smoothly concentrated to their centers, where the star density becomes too great for separating the individual stars, visually or photographically, as shown, for example, in Fig. 48. Short exposures, however, like a 3-minute "snapshot" (Fig. 49) of the Omega Centauri cluster of the southern sky, show the brighter central stars clearly; but when, with the same telescope, we go down to the twentieth magnitude (Fig. 50), seeking stars as faint, intrinsically, as our sun, we "burn out" the center and stop practically all research on the cluster except at the outer edge.

A few globular clusters show in their projected images a slight elongation, indicating in some sectors an excess of stars of perhaps 10 to 15 percent. Omega Centauri is thus elongated, as can be seen by careful examination of the small-scale picture in Fig. 51. The deviation from circularity may indicate the existence of an equa-

Fig. 48. NGC 2419, near the middle of the photograph, is a remote globular cluster over 200,000 light-years away. It is the brightest of the "intergalactic tramp" globular clusters. (Palomar photograph, 200-inch telescope.)

torial bulge produced by rotation around a polar axis that is inclined at a considerable angle to the line of sight. Or it may register the result of past collisions and encounters, such as would be produced by the passage of the cluster through the star strata of our Milky Way. Or perhaps the cluster was born that way. We are still far from the full dynamical explanation of globular clusters.

Messier 13 is popular because its nearness and position in the sky make it available to more than 90 percent of astronomical observers. It can be seen with the unaided eye; and locating it (with the aid of a star map) is one of the interesting exercises for the beginner. Its identification should be in his permanent repertory, along with that of the Andromeda Nebula (Messier 31, Fig. 52), which is the only external galaxy readily visible to northern observers; neither of them is easily discerned, however, except on nonhazy moonless nights.

The amateur's list of naked-eye beacons in the sky should also

Fig. 49. Omega Centauri; a 3-minute exposure which permits the study of the central brighter stars. (Harvard photograph, ADH telescope.)

Fig. 50. Omega Centauri; a 60-minute exposure which in reaching for the faint stars burns out the center. (Harvard photograph, ADH telescope.)

Fig. 51.   Omega Centauri, to illustrate, on a small-scale photograph, the slight elongation that is also shown in Fig. 50. (Harvard photograph, patrol camera.)

include the Orion Nebula (Fig. 2), which is a true nebulosity about 1500 light-years away, and *h* and χ Persei (Fig. 53), a double open cluster in the northern Milky Way. (The Pleiades and the Hyades are open clusters that are too easy.) These four objects are good representatives of four important categories—globular clusters, open clusters, gaseous nebulae, and spiral galaxies—all visible with the unaided eye, and unforgettable after being seen with strong field glasses or small telescopes.

The far-south observer can also see the Orion Nebula and he can use Omega Centauri and 47 Tucanae as outstanding naked-eye globular clusters, the Magellanic Clouds as external galaxies, and Messier 11, or Kappa Crucis (Fig. 54), at the edge of the Coal Sack, for his open cluster. He has available, in fact, much richer fields of stars, and much brighter nebulae and clusters, than we have in the north.

Fig. 52. The Great Andromeda Nebula, Messier 31, and its two small companions. (Palomar photograph, 48-inch telescope.)

Fig. 53.   The double cluster *h* and χ Persei—a pleasant test for the naked eye. (Harvard photograph, 16-inch telescope.)

Among the interesting globular clusters are:

Messier 3, near the pole of the Galaxy in Canes Venatici, distinguished for its many and much-studied cepheid variable stars with periods less than a day (Fig. 35);

Messier 22 in Sagittarius, bright and near, in direction not far from the Galactic center, and situated in the midst of a great star cloud in the Milky Way;

Omega Centauri and 47 Tucanae, conspicuous because of nearness and intrinsic giantism;

Messier 4, in the Scorpion, inconspicuous because of heavy intervening space absorption, though possibly the nearest of all globular clusters;

Messier 62, apparently somewhat malformed;

NGC 2419 (Fig. 48), a globular system found by C. O. Lampland. The cluster was studied by Walter Baade, who found it so distant

Fig. 54. The bright cluster Kappa Crucis, which appears, on small-scale photographs, to dangle into the Coal Sack from an arm of the Southern Cross. The scale of this reflector plate is too large to show either Cross or Sack. (Harvard photograph, Rockefeller telescope.)

(more than 175,000 light-years) that it might be considered not a member of our Galaxy but rather an "intergalactic tramp," or a free and independent citizen of the local group of galaxies. Six of these very faint and remote star systems came to light on photographs made with Schmidt cameras in the 1950s.

Notwithstanding some important differences in stellar composition, the globular clusters are remarkably alike in general appearance. On the basis of their central condensations, however, they are classified into 12 categories. The observed concentrations are to some extent merely a reflection of their distances and the telescopic power used, and not a key to their internal structure. A few clusters, devoid of the usual rich population of giant stars, are what we call "giant-poor"; their globularity and dense population are revealed only when long-exposure photographs bring out the fainter stars. Until then they seem to be merely loose clusters of small population.

Curiously enough, almost all of the now known globular clusters were discovered by Charles Messier and the Herschels (Fig. 55) more than 100 years ago—discovered with small telescopic power as nebulous objects, not recognized as star clusters. The modern large reflecting telescopes had to be used to show that these hazy objects really are star clusters, rather than round nebulae or galaxies; and the reflectors have also been used indispensably to analyze the brighter clusters in detail. Observations of clusters that are revolutionary for theories of stellar evolution have been made in recent years with the 200-inch Hale reflector on Mount Palomar by A. R. Sandage, H. C. Arp, W. A. Baum, and others.

Only a few new globular clusters have been picked up since 1900 in addition to the intergalactic tramps and those photographed in other galaxies. It appears that with the exception of those that lie behind dark obscuration, in the general direction of the galactic center, and except for some of the faint objects in high latitude (probably intergalactic), which are found on Schmidt camera plates, our Galaxy's family of about 100 globular clusters is already known. Photographs in red light will eventually disclose a few of the heavily obscured low-latitude clusters, and large-scale photo-

Fig. 55. A somewhat strange Herschel telescope in a South African field near Cape Town.

graphs may show that some of the rounded images along the borders of the Milky Way are in fact images of globular clusters rather than elliptical galaxies, as now classified. S. van den Bergh speculates that one-third of all globular clusters are intergalactic. E. M. Burbidge and A. R. Sandage have analyzed two of the intergalactic systems and find them to have twice the diameter of typical nearby globular clusters, but to be low in population.

Well before 1920 it was clear that the census of attainable globular clusters in our Galaxy was practically complete and therefore that the whole assemblage could profitably be studied as a system, as a unified aggregation of clusters. Peculiarities came to light at once when the newly estimated distances and the distribution on the sky were examined. It was found, in the first place, that the open Pleiades-like clusters of the Milky Way (galactic clusters, we sometimes call them) were closely concentrated to the galactic circle; they were immersed in rich star fields in all parts of the Milky Way band. Globular clusters, on the other hand, were found chiefly in the southern half of the sky and almost wholly outside the central belt of the Milky Way. This arrangement with respect to the Milky Way circle (the spread, that is, in galactic latitude) is shown in Fig. 56. The globular clusters are found in equal numbers on both sides of the galactic plane; they show a crowding toward the Milky Way, but suddenly disappear just short of its midmost zone.

From the Heliocentric to the Galactocentric Hypothesis

We now know that the globular clusters are much more distant than most of the recorded galactic clusters, and that space absorption contributes strongly to the apparent absence of globular clusters from low galactic longitudes, as are galactic clusters; they are strongly concentrated in the constellations Scorpius, Ophiuchus, and Sagittarius (Fig. 57).

The center of the higher system of globular clusters was found, by Shapley's first analysis of 93 objects then known, to be right on the Milky Way circle in the southern sky, close to the place where the three constellations come together. The right ascension is 17^h 30^m, declination $-30°$. The revised value at the present time differs but little: right ascension 17^h 40^m, declination $-29°5$.

Rather early in the study of globular clusters a somewhat bold

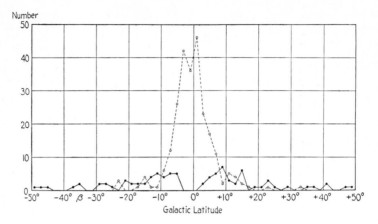

Fig. 56. The contrasted distribution of open clusters (*broken line*), which are crowded into the Milky Way band along the zero of latitude, and of globular clusters (*solid line*), which seem to avoid it. The few recent discoveries of globular clusters would not appreciably alter the graph.

and premature assumption was made. Since the time when the idea was first proposed, however, no one has seriously objected, and many researches on stars, nebulae, and galaxies have tended to remove the assumption from the class of postulates to the class of accepted observations; we have, in fact, lost sight of the original presumption. It was proposed that the globular clusters represent, in a sense, the "bony frame" of the body of the galactic system. It was argued that the spatial arrangement of globular clusters shows the distribution of the billions of galactic stars—shows that the center of the Home Galaxy is in the direction of Sagittarius, for there lies the center of the supersystem of globular clusters.

A new concept of the place of the observer in the stellar universe came as a consequence of these observations and arguments. The heliocentric theory was satisfactory for the planetary system, but no longer sufficient for the stellar system; the sun is no longer to be taken as central among the stars of the Galaxy, but rather as at some tens of thousands of light-years from the galactic nucleus. The fact that the globular clusters are found principally in the Southern Hemisphere comes from the circumstance that they are clustered around the nucleus of the discoidal Galaxy, not around the observer. And it follows that the reason the star clouds of the Milky Way appear somewhat brighter in the direction of Sagittarius and neighboring constellations than elsewhere is that the observer looks toward the rich central nucleus when he turns to that position.

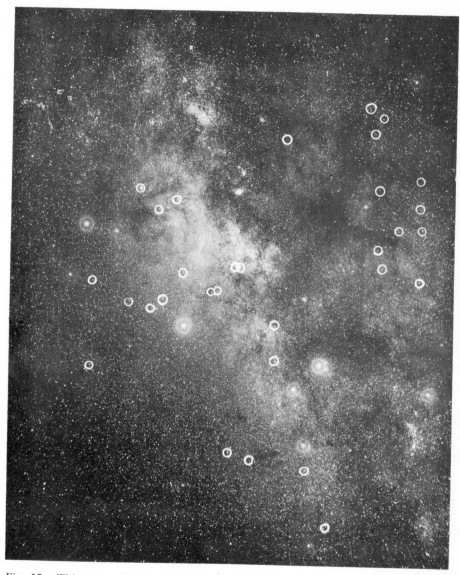

Fig. 57. Thirty-one globular clusters, almost one-third of all known in the galactic system, are shown on a single patrol-telescope photograph of the nucleus of our Galaxy. Circles enclose the cluster images.

An easy picture of the form of the Galaxy, and of our position in it, is obtained from the analogy with an ordinary thin watch. The observer on the earth is located very near the central plane, beneath the second hand; the galactic nucleus, in the Sagittarius direction, is at the center of the watch. We see the Milky Way band of stars when we look in any direction toward the rim of the watch, and more stars, of course (except when obscuring dust clouds intervene), in the direction of the center than elsewhere. When from our eccentric position we look out through the face or the back of the watch, we see relatively few stars; or, in other and more technical words, the star density decreases with increasing galactic latitude. It is, in fact, through the relation of star numbers to galactic latitude that we have deduced the watch-shaped contours of the galactic system.

A comparison with the edge-on spirals lends formidable support to the deduction that our Galaxy is discoidal in form. Around the central axis (of the minute and hour hands, in our analogy), the whole watch-shaped Galaxy rotates, but not as a solid (except near the central axis). The movement of most of the individual stars with respect to the center is probably along somewhat elliptical rather than exactly circular paths.

Details of rotation and internal structure, as far as we have now grasped them, are treated in *The Milky Way*. It is the responsibility of the present volume to look after the general aspects of the Galaxy as seen from outside. But first we should repeat that the hypothesis that globular clusters outline the Galaxy, and locate its center, has been supported thoroughly by work on stellar distribution, stellar dynamics, radio astronomy, and the motions and structural analogues in other spiral galaxies, for example, in NGC 891 (Fig. 58). The measures of radial and transverse motions of neighboring stars indicate that, at our distance from the center (some 30,000 light-years), the speed in the orbit of revolution around the central axis is something like 140 miles per second, and the time required for one trip around—that is, the length of the "cosmic year"—is about 2 million terrestrial centuries. The foregoing numbers will be altered by further research, but they are certainly of the right order of magnitude.

The Thickness of the Galactic System

The analogy of the thin watch gives us the correct impression that the diameter of the Galaxy in its plane is five to ten times the thickness. But we must introduce two important modifications of the analogy. There can be little doubt that the central nucleus

Fig. 58.   NGC 891, in Andromeda, showing its bulged-out central nucleus and the dust clouds along its equator. (Mount Wilson photograph, 60-inch telescope.)

of our Galaxy is elliptical, like that of many spirals (Fig. 9). We should therefore visualize a central bulge covering nearly a quarter of the face and back of the watch. But we do not find as sharp limits to the Galaxy as to the watch. The star population thins out with distance from the nucleus along the Milky Way plane, and also thins out with distance perpendicular to the plane. Analogously, the earth's atmosphere has no sharp boundaries but thins out with height indefinitely. As for the earth's atmosphere, we can gauge our Galaxy's thickness in terms of percentages. For example, we can say that 99 percent of the terrestrial atmosphere is below an altitude of 20 miles, and similarly that 99 percent of the stars in the Galaxy in the sun's vicinity are within 25 light-years of the galactic plane.

The total thickness of the Galaxy, with its centrally bulging nucleus, merits further study. In the all-inclusive system of globular clusters many are at large distances. Some are seen very far above the face of the "watch" and others far below its back. Except for a few very faint and distant clusters recently discovered on Schmidt camera plates, all clusters seem to be physical members of our Galaxy. We are led to wonder if isolated stars also extend out so far. If they do, the over-all shape of the whole Galaxy may not be like that of a watch; it may be spherical, or rather, it may consist of a central discoidal organization like a typical spiral galaxy, surrounded by a roughly globe-shaped "corona" or "haze" of outlying stars.

The existence of a surrounding haze of galactic stars was surmised many years ago, because faint cluster variables had been found in high latitudes—frequently near globular clusters, but apparently not members of them. The reality of this corona has since been definitely established through laborious and systematic studies of faint variables in many latitudes and longitudes. Thousands of stellar photographs, made with several telescopes located at the Cambridge, Bloemfontein, and Oak Ridge (Agassiz) stations of the Harvard Observatory, have been required for the discovery and measurement of the scattered variables. Comparable work was also done in Holland and Germany. The variables of the cluster type (the cepheids with periods less than a day) have been most useful; they have high luminosities (about 150 times that of the sun), and, of greatest importance, they are found all over the sky. The long-period cepheids and the novae, both known as good indicators of distance, are not found in high latitudes and can assist but little in exploring the stellar haze. Moreover, the classical cepheids, as noted earlier, belong to two different populations and serve only when properly identified.

RR Lyrae or cluster-type variables between 30,000 and 50,000 light-years from the plane have now been found on both sides. Moreover, these variables show so definitely a higher frequency as we approach the plane that we can with some confidence conclude that they are all, or nearly all, an organic part of the Galaxy (not intergalactic). It appears, therefore, that the surrounding haze of stars has a total thickness across the galactic plane that approaches, or perhaps exceeds, 100,000 light-years.

The best current value for the diameter of the Milky Way discoid in its own plane is also about 100,000 light-years, a quantity difficult to be precise about because of our awkward inside location and our troubles with the irregular clouds of absorbing dust and gas in low latitudes. Is there, perhaps, a star haze beyond the rim, as well as in high latitude—a haze that increases the over-all dimensions of the Milky Way system in the galactic plane? Have the faint cluster-type variables shown definitely, as suggested above, that the galactic system, because of its haze of stars, is essentially spherical in shape, with a heavy central discoidal structure that contains 99 percent of the stellar mass? Or does this haze of stars so extend in all latitudes that it gives to the surrounding sparsely populated haze the shape of a somewhat flattened spheroid, perhaps twice as extended in the Milky Way plane as at right angles thereto? Observations may eventually demonstrate this last and most likely hypothesis.

Measuring the Boundaries

The size of the neighboring large spiral in Andromeda can be determined without much trouble by making special long-exposure photographs and measuring them with densitometers. There is some difficulty, however, arising from the intervening stars of our own system, which are especiallly annoying when we attempt to measure the exceedingly faint star haze that extends far beyond the visible or ordinarily photographed bounds of the Andromeda system. Nevertheless, the measurement of this corona has been made, photographically and photoelectrically, as well as by the radio telescope. We shall record in the next chapter that the Andromeda system is found to be astonishingly large, in area and in volume, when all the outlying regions are included.

The measurement of the boundaries of our own system, as indicated above, is not so simple. The attack on the problem at the Harvard Observatory has produced, however, preliminary values of the extent of the main discoidal system. Since we are apparently well out toward the rim of the Galaxy, we take advantage of nearness to the boundary in the direction of the constellations Auriga, Taurus, and Gemini to explore that part of the Milky Way and look farther, if possible, into the surrounding haze and the space that lies beyond.

The study of our Galaxy's boundaries, when it involves the use of the numbers, sizes, and brightnesses of external galaxies, must be restricted to the higher latitudes, because the clouds of obscuration fairly well hide from us whatever galaxies there may be in a direction close to the Milky Way circle. In the anticenter direction there is of course heavy space absorption. We have not escaped the darkness by turning away from Sagittarius. We suspect sometimes that there may be around our Galaxy a continuous peripheral ring of obscuration, such as appears to be present in many external galaxies (Fig. 59). There are, however, in the anticenter region, some half-open windows in the obscuration, rather close to the Milky Way circle. Through the thinner dusty haze of these windows many far-distant galaxies can be dimly glimpsed. It is in such semitransparent regions that the study of boundaries can best proceed.

The program for the anticenter area was simple in plan, but tedious in execution. All the sky on both sides of the Milky Way, within about 40° of the anticenter, was profusely photographed with Agassiz Station instruments, which can show variables to the seventeenth or eighteenth magnitude. Some assistance in the program was given with the Cambridge and Bloemfontein telescopes. For each of the 160 separate fields a considerable number of photographs, made on different nights throughout a season, have been intercompared minutely, and changes in the image size of any of the several million stars photographed have been noted. These changing images indicate variable stars, most of them previously unknown. Measures of the photographed images on a large number of plates distinguish the various kinds of variables—eclipsing systems, long-period variables, irregular performers, and some ceph-

Fig. 59. NGC 4594, with its strong peripheral band of obscuration. (Palomar photograph, 200-inch telescope.)

eids. It is these cepheids that we have chiefly sought, since their periods, once derived, indicate at least approximately their absolute magnitudes. There has been much preliminary work with the sequences of standard magnitudes in order to obtain dependable values for the apparent brightnesses of the newly found cepheids. Once we have both apparent and absolute magnitudes, the photometric distances are immediately derived (see Chapter 3).

The measures of the distances of variables in the anticenter region are, however, of little value unless we know that the magnitudes have not been vitiated by space absorption of unknown amount. At this point, therefore, another phase of the research on the anticenter was undertaken. It was possible to judge how much absorption affects one of the variable-star fields by counting for that area the number of external galaxies shown on long-exposure photographs made with the large-field "galaxy hunters"—the Bruce refractor and its successor, the Armagh-Dunsink-Harvard reflector, at the southern Boyden station, and the Metcalf doublet and Jewett reflector at the northern. When to the seventeenth magnitude, for instance, we find as many as 12 galaxies per square degree, we assume either (1) that space is wholly transparent and our computed distances of variable stars in that field will be safe and true, or (2) that the inherent irregularity in the distribution of galaxies has manifested itself in the area we explore, and that in spite of some space absorption (which corrupts the measured distances of the cepheids) the galaxies appear numerous only because an accidentally richly populated region of the universe coincides with our variable-star field. There is nothing much to be done with this unhappy situation, which arises from nonuniformities in the distribution of galaxies, except to smooth out such irregularities, and lessen the consequent errors, by dealing with large areas—by working in many adjacent fields of variable stars and with tens of thousands of galaxies.

The solution can better be found in some modern tricks of photometry. If accurate measures of star brightnesses are made in three or more colors, the anomalous reddening of the stars due to dust can be detected. Then, from calibrations made in nearby star clusters, it is possible to say how much absorption of light must have occurred. In the commonly used UVB system (ultraviolet, visual, and blue), the total extinction in V is just 3.0 times the reddening (the excess in the color $B - V$). To measure the reddening in this way for very faint cluster-type variables is much more difficult and time-consuming than just photographing them; therefore this kind of exploration is proceeding slowly, even though telescopes larger than those at the Harvard Observatory are being brought to bear on them.

It has been assumed without much question (and backed by a good deal of evidence) that our galactic system is a monstrous spiral galaxy of an open type. (The subject has been discussed at length in *The Milky Way*, Chapter 4.) We have shown earlier in this chapter that we are located out toward where the haze begins, or nearly that far from the center of the Galaxy. If photographs similar in power to those we make were made of our galactic system from a suitable location in the Andromeda Nebula, they would show that most of our Galaxy's bright spiral structure is nearer to the center than we are; in fact, it might require very sensitive equipment to record any radiations from our immediate stellar neighborhood. The Andromedan observer should, however, easily photograph a few of our neighboring star clouds, like those in Perseus and Cygnus. The Pleiades should be seen as a faint cluster.

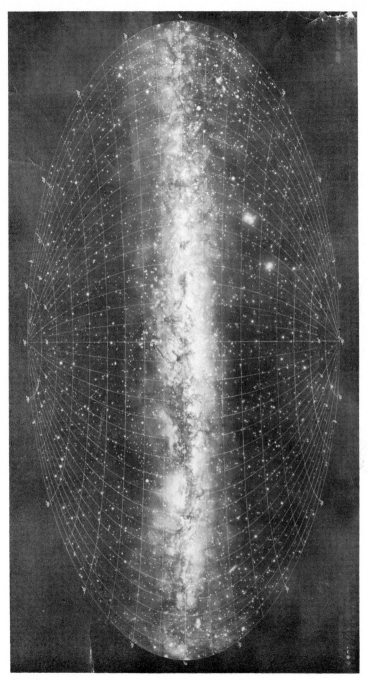

Fig. 60.   Composite picture of the Milky Way. (Lund Observatory.)

Much of the obscuring material that conceals the galactic nucleus is at no great distance, only a few hundred, or a thousand light-years away, and it probably has nothing to do with the nucleus itself. It is in part the dust that lies in the dark lanes along the spiral arms. But wherever the obscuring material may be located, it effectively conceals the nucleus from ordinary view. Figure 60, a composite picture of the Milky Way, shows on a small scale the sum total of light contributed by some thousands of millions of stars. It also indicates how heavily the dark lanes cut into the bright star fields, and how seriously we are handicapped in our researches by the chaotic clouds of obscuring material that shield from our curiosity some of the secrets of the galactic nucleus.

Fortunately, the dust does not shield all radiation from the center. By going to longer wavelengths than our eyes use, astronomers have penetrated the screen of dust and discovered some remarkable things about our Galaxy's nucleus. By means of infrared and radio telescopes it has been possible to map the central region, where several strange bright sources of radiation are found. One of these is the nucleus itself, and the surrounding sources are probably very near it in space. They emit both normal radiation from hot gas and stars and nonthermal radiation, not due to normal stars, but more probably the result of some past explosion in our nucleus, similar to those that are now occurring in radio galaxies and quasars (Chapter 7).

5

The Neighboring Galaxies

Neighborhood is a relative term, and indefinite. It depends on the speed of normal travel and communication and on the size of the total domain. It implies a large nonneighborhood. The earth's persisting neighbor is the moon; comets are only occasional visitors. The neighbors of the sun can be taken as the stars within 50 or 100 light-years, with the billions in the Milky Way excluded. The planets and comets are not the sun's neighbors; they are just constituents of the sun's personal family.

The neighborhood of the Galaxy might be so defined that it includes only the Magellanic Clouds and some vagrant star clusters, all within a radius of 200,000 light-years, or enlarged so that its radius extends to 2 million light-years or so, and then the neighbors would be all the now-recognized members of the local group of galaxies. For the present chapter we choose to take as neighborhood this larger volume. But such a sphere is still relatively very small and leaves out of consideration practically all of the known and explorable universe—leaves out 99.99 percent of it; nevertheless it does encompass about 10^{58} cubic miles.

Fig. 61. A patrol-camera photograph showing Messier 31 and Messier 33, the two nearest spirals. Messier 31 is above the center, to the right; Messier 33, below and to the left, with the second-magnitude star Mirach just halfway between, in the middle of the photograph. The bright star to the far right and below the middle is the highly photogenic Sirrah (α Andromedae), which is also of the second magnitude. (Oak Ridge photograph by Henry A. Sawyer.)

The Andromeda Group

The neighborliness of the Magellanic Clouds, and their useful cooperation in the task of exploring galaxies, has been recorded in two of the preceding chapters. The Andromeda group of galaxies, we find, is not as conveniently located when we want to borrow astronomical tools and obtain general cosmic information. The nearly 15 times greater distance of those neighbors conceals from us some of their inner secrets that would be useful if revealed, and perhaps would be easy to learn if they, too, were but 160,000

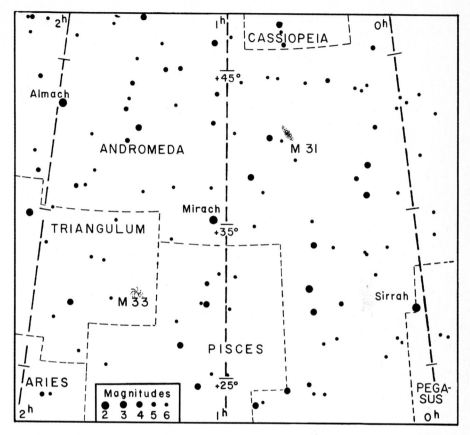

Fig. 62. A diagram showing the location of the Andromeda Nebula and Messier 33 among the naked-eye northern stars.

light-years away. Nevertheless, this neighborly group should be credited with leading us directly to basic knowledge of the outlying universe. Stated otherwise, the archipelago in Andromeda has provided preliminary steppingstones for our plunge into the distant oceans of space and time, where with some success we now grasp at thousands of other "island universes," and glimpse a billion more on the cosmic horizons.

The Andromeda Galaxy

The most conspicuous spiral galaxy in the sky is the great nebula in Andromeda (Fig. 52), also known as Messier 31 or as NGC 224 (Figs. 61 and 62). It is a very large and bright object with sufficient luminosity that it can be seen on a good night even without a

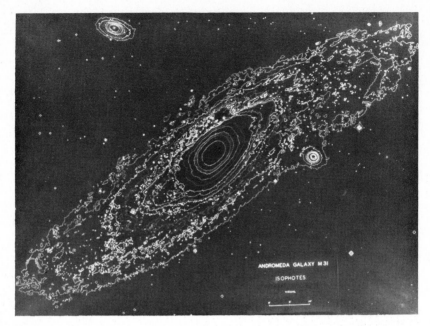

Fig. 63.   An isophotal diagram of Messier 31, the Andromeda galaxy, determined by establishing the location of lines of equal brightness (compare with Fig. 52). (From K. Krienke.)

telescope. Long-exposure photographs show that its longest dimension is at least 4° across the sky, eight times the size of the full moon. Its absolute luminosity, calculated on the basis of its distance, is also found to be very high, making it in fact one of the brightest spirals in nearby space.

In structure the Andromeda nebula is composite, as is generally the case for spiral galaxies. In its central area it is found to have a distribution of luminosity that is very much like that of globular star clusters or elliptical galaxies (Fig. 63). It is very smooth and steadily decreases outward according to a specific mathematical law. In the outer parts, where the conspicuous spiral-arm structure shows up, an additional component contributes to its structure; this is the so-called "disk" part that appears to be flat and that has a structure which is mathematically of a form different from that of the inner parts. The disk component is apparently very flat and it is here that the conspicuous bright stars, gas clouds, dust lanes, and neutral hydrogen show up. About 75 percent of the total light of the Andromeda galaxy comes from the disk component, and the rest from the superimposed elliptical-galaxy component. Measures of the light intensity taken across the Andromeda galaxy (Fig. 64) show that the spiral arms, which appear so conspicuous on photographs, are actually rather inconspicuous bumps on the luminosity curve and contribute only a modest fraction of the total light of the galaxy.

In color the nebula is different from one place to another. Color photographs taken at Mount Palomar show clearly that the central area is yellowish and the outer parts, in the disk component, are more bluish in color. Precise measurements of the color by means of photoelectric photometers indicate that in the central area, where the yellow color shows up, the light appears to come primarily from old, evolved giant stars, not too different from those found in the oldest clusters (galactic and globular) in the Milky Way. The spiral arms are very blue and their colors can be quantitively understood as being due to the fact that they contain large numbers of very young stars (Fig. 65). The reddest portions of the Andromeda galaxy are those that are heavily obscured by dust clouds, and the redness is therefore interpreted as being due to reddening by this intervening dust. Especially just west of the central region, there are very conspicuous dust lanes where the color is measured to be extremely red.

The Andromeda galaxy is inclined very steeply to the apparent plane of the sky so that it makes an angle with the line of sight of only 12°. It is therefore seen almost edge-on and as a result it is difficult to untangle the spiral-arm structure, especially along the minor axis where the arms are almost superimposed on one another as we see them. Nevertheless, by restricting attention to the large gas clouds or by examining particularly the bright stellar associations it is possible to delineate these spiral arms and to discover

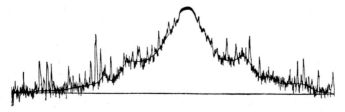

Fig. 64. Densitometer tracing along the major axis of Messier 31. Field stars and plate graininess contribute to the irregularities; the undifferentiated starlight within Messier 31 provides the larger trends in the tracing.

Fig. 65. Supergiant stars involved in the dust and gas of the outer whorls of the Andromeda Nebula. (Mount Wilson photograph, 100-inch telescope.

that the galaxy contains two main arms that make several trips around the center as they spiral outward.

The distance to the galaxy is found primarily by measuring the cepheid variables and the novae. Hundreds of cepheid variables are now known, owing to the discoveries made initially by Hubble (Fig. 66) and more recently by Miss Swope, Baade, and Gaposhkin. The period-luminosity relation for some of these cepheid variables is shown in Fig. 67 and it is found to be virtually identical in shape and characteristics to that for cepheid variables in our Galaxy and in the Magellanic Clouds. Therefore there is little reason to doubt that the cepheids provide an accurate means of measuring the distance to the Andromeda galaxy, which is found to be 2.2 million light-years.

There are many novae observed in the Andromeda galaxy, some 50 being observed in one year in a program of search for these objects. These also can be used to measure the distance to Andromeda by comparing their luminosities and their rates of decline of luminosity after maximum with those of similar objects in our Galaxy for which distances can be measured by other means. The result is almost identical to the result of measuring the distance by means of cepheid variables. Therefore we have considerable

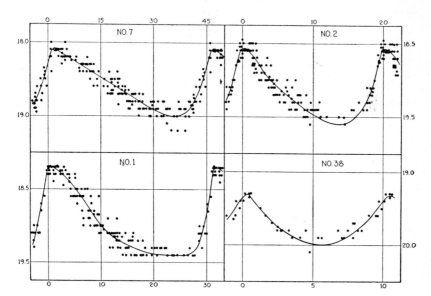

Fig. 66.   Light curves of cepheid variables in the Andromeda Nebula, as derived by Hubble at Mount Wilson.

M 3 1

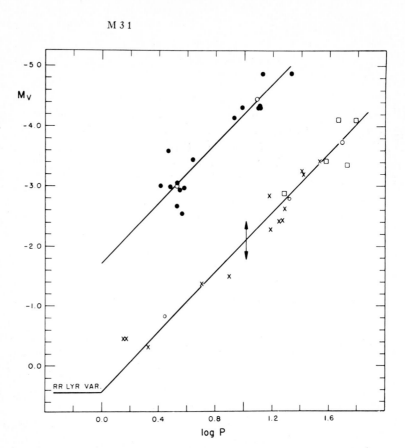

Fig. 67. The period-luminosity relation for variable stars in an out-lying region of Messier 31. The upper curve is for the normal cepheid variables, and the lower curve is for the Population II cepheids. (From Baade and Swope, *Astronomical Journal*.)

confidence that we know now the distance to the Andromeda galaxy to within about 10 percent.

The total mass of the Andromeda galaxy is very large. Recent measurements of its rotation and the way in which the rotation varies from the center outward to the outermost parts have allowed astronomers to make reliable estimates of this total mass. This has been done both by means of the 21-cm radio measures of the neutral hydrogen emission and by means of optical measures of the velocities of the gas clouds seen in the Andromeda galaxy's spiral arms. In both cases the mass is found to be approximately 300 billion times the mass of the sun, with an uncertainty of 25–30 percent.

The mass of gas in the galaxy is found to be 2 percent of this total, the rest of the mass being primarily in the form of stars.

The nucleus of the Andromeda galaxy is small, very bright, and reddish. It is slightly elliptical in outline with a diameter of about 300 light-years. It is rotating very rapidly, much more rapidly than the surrounding portions of the galaxy, for reasons that are not yet understood.

Star clusters and stellar associations in the Andromeda galaxy have been studied to see if they can give us any information on the nature of the stellar population. The globular star clusters were originally studied by Hubble, who catalogued them, and in more detail in recent years by Hiltner, Kron, and van den Bergh. In many respects they are very similar to the globular clusters in our Galaxy, but with some very strange differences (Fig. 69). For example, although in our Galaxy most globular star clusters at large distances from the center are very poor in heavy elements (because,

Fig. 68. The 36-foot radio telescope operated by the National Radio Astronomical Observatory on Kitt Peak, Arizona. (Kitt Peak National Observatory photograph.)

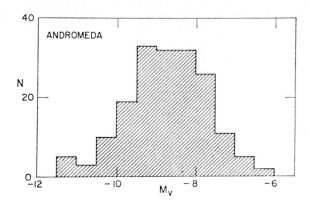

Fig. 69. The luminosity function (distribution of luminosities) for the globular star clusters of Messier 31. (After van den Bergh.)

we believe, they were formed at an early epoch in the history of our Galaxy, when heavy elements had not yet been formed in the nuclei of stars), in the Andromeda galaxy the situation is different. Many of the globular star clusters, even some at very large distances from the center, are relatively rich in heavy elements. Their compositions are more like that of the sun than like that of the distant globular clusters in the Milky Way halo. We do not know the explanation for this nor do we see at the moment any way in which this can be accounted for in terms of differences in the early histories of the two galaxies. Because most of these globular clusters are so far away and so compact, there is little hope of studying them much more completely than we have now until a much larger telescope with better resolving power than the largest telescope now in use (the 200-inch reflector at Palomar Mountain) is constructed. Therefore this particular mystery, the strange chemical abundances in the globular clusters of the Andromeda Nebula, may remain for many decades.

The stellar associations, which are large loose groupings of very young stars, are conspicuous in the spiral arms of the Andromeda Nebula. Van den Bergh has studied nearly 200 of these and found them to be rather large (up to almost 3,000 light-years in diameter) and very blue. The nearest to the center are 20,000 light-years out and the most distant catalogued is 80,000 light-years from the center. The sample of stellar associations in Andromeda, like that in the Magellanic Clouds, allows us to study these objects in a much more comprehensive way than we can in our own Galaxy, where these associations are mixed up so completely and confusingly with dust clouds and intervening extinction and absorption. It is possible from the study of these objects to see the way in which they evolve, because the ones that turn out to be more clumpy contain the youngest stars, whereas the others, which are more smoothly structured and systematically larger, contain somewhat older stars. This suggests that such groupings of stars evolve from clumpy, bright, blue, small groups to expanded and more evenly populated groups.

Not only do we have considerable information from radio telescopes on the distribution and amount of the neutral hydrogen gas in the Andromeda galaxy (Fig. 70) but we also find from radio telescopes that there is another kind of radio radiation emitted by this galaxy. It is rather weak but easily detectable with the giant radio telescopes now in operation. About one-tenth of this so-called "continuum" radiation comes from the spiral-arm structure and is believed, by analogy with our own Galaxy, to be emitted by the very hot gas clouds that we can see in the arms. The rest comes

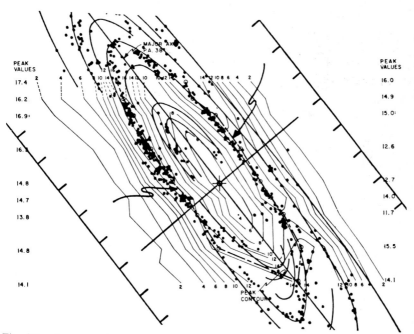

Fig. 70. A comparison of the neutral-hydrogen density (*fine lines*) with the distribution of optically bright gas clouds (*dots*) in the Andromeda galaxy. (From M. S. Roberts, *Astrophysical Journal*, copyright University of Chicago Press.)

from a much larger area forming a halolike envelope around the galaxy. This may come from a number of small discrete sources, possibly supernova remnants, or it may come from a general diffuse halo of high-energy particles.

Fifty Million Andromedan Novae on the Way

The behavior of violently disturbed stars, such as the novae, can be studied in many nearby galaxies, thereby increasing greatly our knowledge of their characteristics. A typical light curve is shown in Fig. 71. From an extensive survey of the Andromeda Nebula, Arp studied many such nova light curves. It is interesting to note that if the present rate of "novation" holds for this nebula—and we see no reason why it should not—there has been in the past 2 million years much as yet unrecorded stellar violence. The complete records of the light eruptions, radial motions, and spectrum changes for about 50 million novae are in the light waves on their way from the Andromeda Nebula to the earth—phenomena of the past for that evolving galaxy, of the future for us.

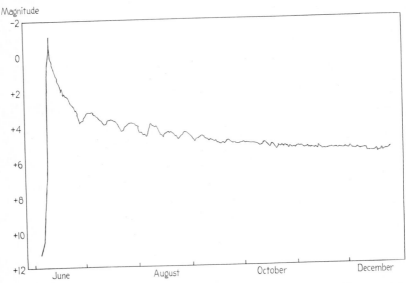

Fig. 71. Light curve of Nova Aquilae, 1918, the brightest nova discovered in our Galaxy since Kepler's supernova of 1604.

The Companions to the Andromeda Nebula

The Magellanic Clouds are well established as close companions to our Galaxy. Similarly, the Andromeda Nebula is attended by companions, but in its case there are four and all four are elliptical galaxies. Two of the companions, NGC 205 and M 32, are very close to M 31, the Andromeda galaxy; the other two, NGC 147 and NGC 185, are considerably more distant.

The two more distant galaxies are separated in the sky from M 31 by about 6°. From the currently accepted distance for these objects, about 2,000,000 light-years, this means that their actual physical distance from M 31 is approximately 450,000 light-years. Therefore, they can hardly be considered companions in the same sense as the Magellanic Clouds, which are only 160,000 light-years from our Galaxy. They nevertheless are probably physically associated with the Andromeda Nebula. The evidence of this association includes the fact that their radial velocities with respect to the sun are virtually the same as that of M 31. Furthermore, there is evidence that the outer structures of these two companion galaxies are dominated by the tidal effect on their outer stars caused by the Andromeda Nebula. NGC 147 and NGC 185 are only 1° apart and therefore are close in space and possibly rotate around each other, although there is no evidence yet to that effect.

NGC 147 is the faintest and least conspicuous of the four companions to the Andromeda Nebula. It barely shows up on small-scale photographs and is an extremely difficult object to photograph with anything but the largest telescopes. It is highly elongated and its luminosity is mildly concentrated toward its center. There is no clear nucleus, just a faint patch near its center that may actually be only a globular star cluster that happened to be projected on the center of the system from our viewpoint. In color it is very uniform throughout and the color is consistent with its being made up primarily of stars very similar in their properties to those of globular star clusters.

The actual stellar content of NGC 147 first came to light to 1944 when Walter Baade first resolved the galaxy into individual stars with the Mount Wilson 100-inch telescope. He estimated that the brightest stars were of about the same magnitude as the brightest stars in the closer companions to the Andromeda Nebula (and also virtually the same as the brightest stars in the nucleus of M 31), and from that concluded that the distance to NGC 147 must be about the same as the distance to M 31. Because of the extreme difficulty of photographing these brightest stars even with the world's largest telescopes, no one has improved upon Baade's estimate of its distance. It takes a night of exceptionally steady atmosphere and dark sky to register the faint images photographically and no one has attempted the much more difficult task of measuring these stars photoelectrically. They are at the faint limit of present photoelectric techniques and are crowded together so thickly that it would be virtually an impossible job to center a photometer on an individual star and register its brightness. For that reason, we have only a very approximate estimate of the brightnesses of the brightest stars and therefore of the distance to this galaxy.

Though its distance is almost impossible to determine, it has been possible to measure accurately its total luminosity. The Swedish student of galaxies, Erik Holmberg, measured its total brightness photographically in 1958 and found it to have a visual magnitude of 9.7. If it is truly at the same distance from us as the Andromeda galaxy, this means that its absolute magnitude is -14.8. It is therefore more than 100 times fainter than M 31. NGC 147 contains a few globular star clusters, the colors of which seem to be normal.

The brighter of the two distant companions, NGC 185 (Fig. 72) has a total brightness of $M_V = 9.4$. It is smaller than NGC 147 and more compact. Its center is brighter and is of more circular outline, though it has no apparent nucleus. In shape it is very nearly perfectly elliptical, but with a noticeably higher ellipticity in the center than in the outer regions. The ratio of the minor to the major axis of lines of equal intensity changes outward, going from a value of 0.81 in the center to a value of about 0.87 at a distance of about 1° from the center. Like NGC 147, NGC 185 contains a few globular star clusters which appear to have normal colors. The color of the main galaxy, however, is not normal, apparently owing to the presence in its innermost regions of a number of bright blue stars. These were first discovered by Baade and only recently have been studied in any detail. It is found that of the more than 50 bright blue stars found in NGC 185, all appear to be normal main-sequence stars, unusual only because they are much too bright to be on the main sequence of a Population II object like an elliptical

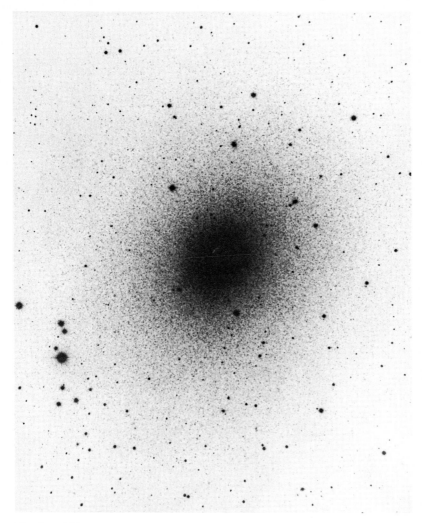

Fig. 72.   NGC 185, a distant companion to Messier 31, in a positive print. (Palomar photograph, 200-inch telescope.)

galaxy. It is not yet known what their presence indicates, other than that it is apparently true that star formation on a small scale has occurred in at least two epochs for NGC 185. The first epoch includes the formation of the vast majority of the stars, which now make up the faint reddish globular-cluster-like portion of the galaxy, and the more recent epoch has contributed the few bright blue stars near the center. Of two alternative explanations of these

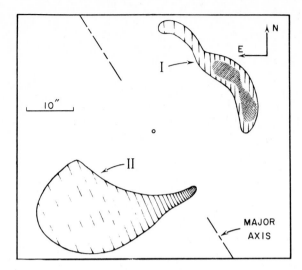

Fig. 73.   Sketch showing the location of the two dust clouds (dark nebulae) near the center of NGC 185.

objects, one is that they have formed from nonstellar material that has been ejected over the recent millennium by the evolving low-mass giant stars, and another is that they have been formed by the gas and dust that may have been dispersed by a mild explosive event in the center of NGC 185, similar to those that are observed to occur in centers of large radio galaxies. There are two dust clouds associated with the central region of NGC 185 (Fig. 73) and these are probably connected in a generic way with the bright blue stars. They seem to be approximately symetrically distributed about the center at a distance of about 15″ of arc (between their inner edge and the center). The dust clouds are in average dimension about 100 light-years across and have total estimated masses of 150 and 20 times the mass of the sun, respectively.

The two elliptical galaxies that lie closest to the Andromeda galaxy are both brighter than the more distant companions. NGC 205 (Fig. 74) is highly elongated and is positioned, at least as we see it, among the outermost stars of the Andromeda system. On long-exposure plates taken with very large telescopes it is not possible to discover where the Andromeda galaxy stars stop and the stars of NGC 205 begin; they appear to melt together. In many respects NGC 205 is similar to its somewhat fainter companion, NGC 185. About a magnitude brighter in total luminosity and

Fig. 74. NGC 205, companion to Messier 31. (Palomar photograph, 200-inch telescope.)

somewhat larger, it contains a collection of bright blue stars that is very similar to that in NGC 185. Also, dust clouds are conspicuous in its center, and the effect of this anomalous grouping of Population I objects is easily measurable in the color and the color dis-

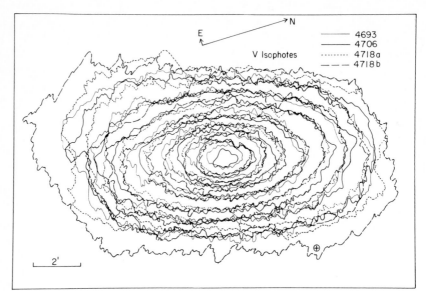

Fig. 75.   Sets of isophotes for NGC 205, showing lines of equal luminosity. (From Hodge, *Astronomical Journal.*)

tribution for this galaxy (Fig. 75). The outer portions of the galaxy show a distortion in their shape that is probably due to the closeness of the Andromeda galaxy. The spectrum of NGC 205 has been examined with a photoelectric spectrum scanner by the Berkeley astronomer Hyron Spinrad, who has been able to show that the amount of energy received in the different color regions of the spectrum gives information on the kinds of stars and on their chemical composition. He finds for NGC 205 that the spectrum is very similar to the spectra of certain globular star clusters of our Galaxy, like M 5, which contain fewer of the heavier elements than are found in the sun.

The companion that has apparently the smallest distance from the Andromeda Nebula is the bright compact system M 32. It is about equal in total luminosity to NGC 205, but because of its much smaller size it is a more conspicuous object. It has a bright nucleus that is distinctly elongated and very small, less than 1″ of arc in diameter. The Lick astronomer Merle Walker has been able to measure the radial velocities of this nucleus at different distances from the center and has found that the nucleus rotates very nearly like a solid body, with a maximum velocity of rotation of about 45 miles per second at its edge. Spectrum scans by van den Bergh and Henry have shown that M 32 is also very much like a globular star cluster in the composition of its stars. The total mass of the galaxy can be estimated from the velocities found in its central regions, which show a dispersion of about 60 miles per second. This has led to the conclusion by the Burbidges (a University of California astronomical husband and wife team) and by Ivan King that the total mass must be about 4 billion times the mass of the sun.

Messier 33

Another galaxy that might be included with the Andromeda archipelago is the small spiral galaxy M 33 in the constellation Triangulum (Fig. 76). It contains many of the same kinds of stars as are identified in M 31, but our information on these stars is still sketchy because of the lack of very much published photoelectric work. Messier 33 is an open-armed *Sc* spiral and is notable for its richness in supergiant blue and red stars, gas clouds, dust lanes, and star clusters. Hubble's original detailed examination of its stellar content at Mount Wilson in the 1920s showed that its distance could be determined from the numerous cepheid variables within it. He found it to be only slightly different from the distance to the Andromeda Nebula. In addition to the cepheids he also found novae and a few bright blue erratic variables that behaved something like S Doradus in the Large Magellanic Cloud. Similar stars have also been found in the Andromeda galaxy and a few other nearby spirals.

Among the objects measured and studied in detail in M 33 are large numbers of faint clusters. Apparently all of the clusters identified so far in M 33 are galactic clusters; not a single one of them is clearly a globular star cluster. Therefore, it may be true that M 33 is even more of a chiefly Population I object than we might claim for the Magellanic Clouds, which at least have a few normal globular star clusters. Perhaps the difference is not real, however, and the clusters in M 33 have simply not been adequately studied yet.

Messier 33 contains many stellar associations and bright gas clouds. Although the stellar associations and other supergiant stars have not yet been studied in detail, the brightest of this galaxy's

Fig. 76. Messier 33, a long-exposure photograph showing gas clouds, dust clouds, and supergiant stars. (U. S. Naval Observatory Flagstaff Station photograph by John Priser, 61-inch telescope.)

gas clouds, NGC 604, has been analyzed by a number of astronomers. Determinations of its chemical composition have shown that the abundances of the elements are very similar to the abundances found for the gas clouds in our Galaxy.

The total mass of the M 33 spiral has been estimated both from radio observations of its neutral hydrogen and from optical observations of its gas clouds. J. C. Brandt in the United States and Courtes and his collaborators in France have made the most comprehensive studies of the velocities of the gas clouds visible by optical means. Kurtiss Gordon, using the 300-foot radio telescope at the United States National Radio Observatory in West Virginia, has made a study of the neutral hydrogen in this galaxy. The total mass of neutral hydrogen is about 10^9 times the mass of the sun and the total mass of the entire galaxy is approximately 30 times greater.

Two Dwarf Irregular Neighbors

Two more members of the local groups of galaxies are the small irregular systems known as NGC 6822 and IC 1613. The second was discovered too late to be included in the massive *New General Catalogue* assembled by Dreyer of the Armagh Observatory in Ireland, and so he placed it along with hundreds of other uncatalogued objects in one of two *Index Catalogues*. Hence, the letters IC.

These two galaxies are both irregular in form like the Magellanic Clouds. They are dwarfs because they are both small in extent and faint. Their total luminosities are each less than the luminosity of NGC 205 or M 32. Their irregular structure and low velocities are important in our consideration of the makeup of the universe. Four of the 14 members of the local group of galaxies that we have discussed so far are small irregular galaxies, apparently misshapen and confused aggregations of stars. In the general surveys of the thousand or so brightest galaxies, on the other hand, only about 4 percent are found to be irregular in form. Of the thousand apparently brightest galaxies, about 75 percent are spirals and a little over 20 percent are ellipticals.

Why this difference for irregulars—28 percent in one case and only 4 percent in the other? Is our part of the universe not typical of the whole? Is there something defective in our surveys? Are the laws of chance playing tricks?

The trouble is obviously with the surveys. The dwarf irregular galaxies are recorded only if they are near, while the giants are recorded over much greater distances. Therefore the catalogues of the brightest galaxies favor the giants and these giants in luminosity are preferentially spirals and ellipticals.

If NGC 6822 and IC 1613 were located 50,000,000 light-years away instead of about 2,000,000, our best photographs might not show them at all. Also, if they were near the limit of faintness of our photographic plates, we might fail to see their irregularity and might class them with the faint elliptical types or mistake irregular extensions as spiral arms and put them down as spiral galaxies.

As an example of how such small irregular galaxies are sometimes mistakenly identified as elliptical galaxies, Fig. 77 shows a photograph of a faint galaxy in the nearby Leo group, near NGC 3377, a giant elliptical galaxy. In photographs taken with the 48-inch Schmidt telescope at Palomar Mountain, the dwarf galaxy looks like a faint elliptical galaxy like NGC 147 and was originally classified as such. But in Fig. 77 from a plate taken with the 120-inch telescope at the Lick Observatory, it is clear that this dwarf galaxy is resolved into individual blue supergiant stars, and that it has an irregular form not unlike that of IC 1613.

So far, at least, in our tally of the numbers of different kinds of galaxies in our local neighborhood we have found six extremely faint dwarf elliptical galaxies, four moderately faint elliptical galaxies, four small irregular galaxies, two giant spirals, and a moderately bright spiral. Does this represent a typical distribution of galaxies in the universe? We do not know the answer because of the great difficulty of sampling dwarf galaxies in distant reaches of space. For example, the very faint elliptical galaxies like the systems in Sculptor and Fornax would be extremely difficult to detect even if they were as close as in the neighborhood of the Andromeda Nebula. We do have surveys which show that in some of the nearby large clusters of galaxies dwarf elliptical galaxies do exist, but so far we have not yet detected any as faint as the extremely faint ones in our immediate neighborhood.

NGC 6822. E. E. Barnard was an inspired amateur astronomer living in Nashville, Tennessee, when in 1884 his small telescope picked up a faint nebulous patch not far from the southern Milky Way. At first it appeared to be made up of several disconnected nebulosities, but when photographs were taken of it at Mount Wilson it was found to be one galaxy with several bright condensations and gas clouds within it (Fig. 78). This is the object that became number 6822 in Dreyer's *New General Catalogue*. It is a difficult object and requires considerable telescopic power to show the mixture of

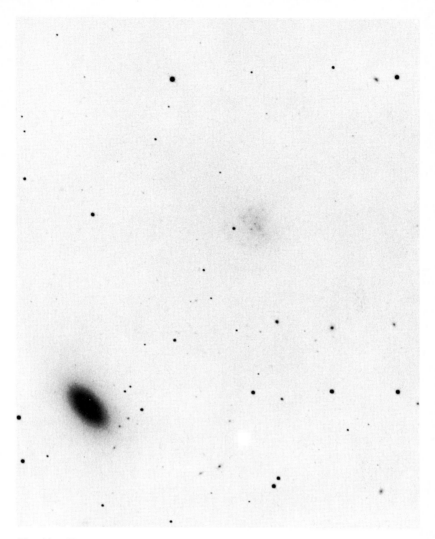

Fig. 77.   Two galaxies in the nearby Leo group. In the lower left-hand corner is the bright elliptical galaxy NGC 3377 and just above and to the right of the center of the photograph is a faint dwarf irregular system (uncatalogued). (Lick photograph, 120-inch telescope.)

stars and nebulosity together. Hubble pointed out, after studying it with the large telescopes at Mount Wilson, that a low-power eyepiece on a 4-inch telescope shows the object as fairly conspicuous but it is barely discernible at the primary focus of the 100-inch telescope. The former concentrates the light and the latter spreads it out.

Fig. 78. Barnard's galaxy, NGC 6822, faintly shown through a southern star field. (Mount Wilson photograph by Hubble, 60-inch telescope.)

Lying in the constellation of Sagittarius, Barnard's galaxy was one of the first galaxies to be studied in great detail with the giant telescopes at Mount Wilson. In his first paper dealing with the nearby galaxies and their stellar content, Hubble showed that NGC 6822 must be a remote stellar system. He was able to determine its distance and size because he found in it 11 cepheid variables from which the distance could be estimated by means of the period-luminosity relation. The Milky Way is only 20° away and as a consequence there are many field stars superimposed on the galaxy. Nevertheless, it was clear from Hubble's study that NGC 6822 in many ways resembles the Magellanic Clouds. Its cepheids are typical in all of their properties.

Until recently very little research was done on NGC 6822 after Hubble published his first paper on the system in 1925. An important new development in its study occured in the middle 1960s, when several astronomers turned their attention to this galaxy. The first was Susan Kayser, a graduate student at California Institute of Technology, who, in searching for a thesis subject, hit upon the

idea of redoing Hubble's study of NGC 6822 by modern photometric methods. Her thesis advisor, Halton Arp, obtained photoelectric measures in the vicinity of the galaxy down to very faint
limits and supplied her with Hubble's plates as well as many
subsequently obtained photographs of this object. In a comprehensive study of the stellar content, including the cepheid variables,
of which she discovered several in addition to Hubble's original
list, she was able greatly to improve upon our knowledge of the
properties of this remote system. Figure 79 shows the period-
luminosity relation from her study and Fig. 80 gives the color-

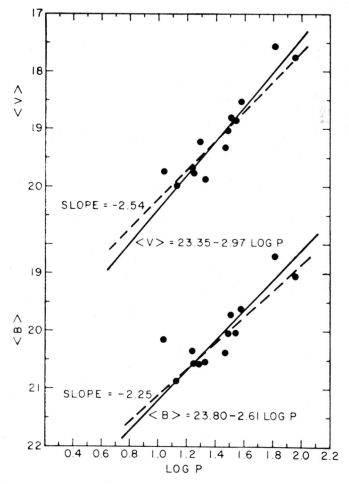

Fig. 79. Period-luminosity diagrams for cepheid variables in NGC 6822. The
upper diagram is for yellow light (V) and the lower for blue light (B). (From
Kayser.)

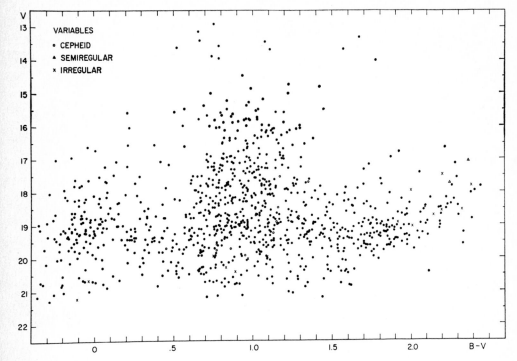

Fig. 80. A color-magnitude diagram for the brightest stars in NGC 6822. The variable stars are identified. Most of the stars in the center of the diagram, of intermediate color, are foreground stars. (After Kayser.)

magnitude diagram that she obtained by a photographic survey of measurable stars. The latter figure shows the heavy contamination by foreground stars of our own Galaxy, a consequence of the low galactic latitude of this object.

A recent study of NGC 6822 by Hodge involved the photoelectric and photographic determination of its structure. Figure 81 shows the way in which the stars in this system are arranged, and Fig. 82 gives some photometric isophotes for it. It is clearly very similar in its structure to the Large Magellanic Cloud, with the exception that only the bar portion of the system is heavily populated with stars. At the north side of the bar there are several bright gas clouds (H II regions, astronomers call them), and study of the neutral hydrogen by radio telescopes shows that the neutral gas of the galaxy extends over a very much larger area than the discernible star component.

A recent determination of the chemical composition of the galaxy

λλ4900-5700A

103a-D EMULSION
PLUS GG-11 FILTER

RECORDS EMISSION IN
N_1 AND N_2 [OIII]

λλ5100-5700A

103a-D EMULSION
PLUS GG-14 FILTER

RECORDS NO EMISSION

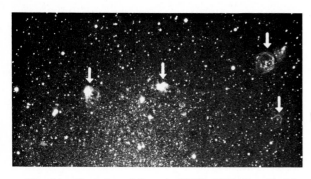

λλ6400-6700A

103a-E EMULSION
PLUS RG-2 FILTER

RECORDS EMISSION IN Hα

Fig. 81.   Emission objects in NGC 6822 identified by means of various plate
and filter combinations. (Mount Wilson and Palomar photographs.)

NGC 6822 results from a study of one of the bright gas clouds visible
in Fig. 78. Manuel Peimbert of the Instituto de Astronomia, Uni-
versidad de Mexico, and the Observatorio de Tonantzintla in
Mexico and Hyron Spinrad of Berkeley made photoelectric scans

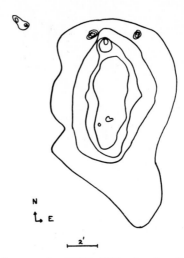

Fig. 82. Isophotal diagram for NGC 6822, showing lines of equal intensity of light. The emission regions are visible as peaks near the top of the diagram. (From Hodge, *Astronomical Journal.*)

of the emission lines in one of NGC 6822's gas clouds with the 120-inch telescope at the Lick Observatory. They concluded that the galaxy is deficient in nitrogen and oxygen as compared to our own Galaxy, a conclusion based on a direct comparison between the gas cloud there and two bright gas clouds near the sun, in Orion and in Sagittarius. Nitrogen is less abundant in NGC 6822 by a factor of 6 and oxygen less abundant by almost a factor of 2. Helium, on the other hand, is about the same in its relative abundance to hydrogen as we find in our own local Galaxy. The surprising difference for the heavier elements has profound significance with regard to arguments about both the evolution of the elements in the universe and the principle of uniformity that is so important to cosmological explorations. In the first instance, the lack of the heavy elements in some parts of the universe suggests that these objects cannot have been formed in any primeval event but must instead be formed primarily in stars, to a varying extent in different environments. On the other hand, the nearly exact correlation in the amount of helium found in different galaxies suggests that the helium could be primeval, at least in part, owing to element formation near the beginning of the universe.

The distance of Barnard's galaxy is almost the same as the distance to the Andromeda group. Its spectrum shows a small blue shift, indicating that relatively the observer and NGC 6822 are approaching each other; the speed is about 22 miles per second, according to Lick and Palomar measures. But this apparent approach comes from the observer's rotational speed in his own Galaxy. The centers of our Galaxy and NGC 6822 are actually receding from each other at the rate of 60 miles per second.

Unlike the Magellanic Clouds, this more distant dwarf galaxy is scarcely in the satellite class, for it lies far beyond the most remote stars yet found in our Galaxy's surrounding star haze.

IC 1613. The second of the small irregular galaxies in our local group is the one known as IC 1613. This galaxy might well be called Baade's galaxy, because Walter Baade spent so much of his astronomical life in its study. Its total absolute magnitude is almost identical to that for NGC 6822, but IC 1613 is about twice as big in linear extent. It is more spread out and less complex in structure than Barnard's galaxy. It is somewhat similar to the Magellanic Clouds, but much less complex, smaller and fainter.

In appearance IC 1613 is remarkably simple and uncomplicated. This has allowed some of us to select it as a prime galaxy for the study in detail of some of the ways in which galaxies evolve and change in time. It has no spiral structure, it has no barred structure, it has a wide variety of different kinds and ages of stars, it is small and limited in extent, and it is close to us. These features make it an extremely good object for testing out our ideas of patterns of star formation in galaxies. Figure 83 shows a photograph of IC 1613 taken with the 100-inch telescope at the Mount Wilson Observatory, illustrating the relative simplicity of it. The most conspicuous feature is a bright association of stars in one corner, which contains not only the only gas clouds in the galaxy but also the brightest stars and the longest-period cepheid. This is the only area in the entire galaxy where there is evidence that star formation is now going on. Other star groupings can be made out in other parts of the galaxy and these are clearly groupings with greater ages than the bright one. Attempts to reconstruct the recent history of this galaxy have been made by age-dating the different star groups that seem to make up most of the irregular structure of IC 1613 and these attempts have successfully shown the way in which the galaxy has changed with time (Fig. 84). We can trace the areas which at various times in the past were bright and active and can in this way discover the complicated pattern by which the galaxy has developed.

Among the many remarkable features of this small galaxy is the

Fig. 83.   IC 1613, an irregular neighbor that is rich in cepheid variables. (Mount Wilson photograph, 100-inch telescope.)

fact that no one has ever discovered a single star cluster in it. This is all the more remarkable when we realize that it has one-tenth the total brightness of the Large Magellanic Cloud, and therefore probably contains about one-tenth as many stars, and yet the Large Magellanic Cloud is estimated to contain on the order of 6000 star clusters. We would predict, therefore, that at least a few hundred

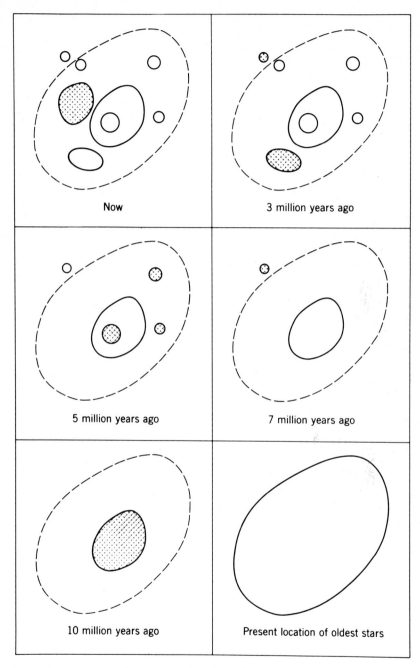

Now

3 million years ago

5 million years ago

7 million years ago

10 million years ago

Present location of oldest stars

Fig. 84.   Diagrams showing the location of areas of star formation at different epochs during the recent history of the galaxy IC 1613. Cross-hatched areas indicate locations of intense star formation activity. (From P. W. Hodge, *Galaxies and Cosmology,* courtesy McGraw-Hill Book Company.)

star clusters should show up on our best photographs of IC 1613, but not a single one does. This is a complete mystery; no one knows what determines the number of star clusters that form in a galaxy, nor what effects might inhibit their formation or promote their disruption in such a complete way as to lead to this huge discrepancy.

The variable stars of IC 1613 have been studied by both Baade and Sandage. Most of the variables that have been studied in detail so far are cepheids with periods ranging from a few to more than 100 days. There are several irregular variables and one long-period variable. There seems to be only one eclipsing binary discovered so far, as well as one nova and one very remarkable star that Baade was completely unable to understand. This latter has a period of 29 days with what Baade described as an apparently upside down light curve. It may be a variable of a new type, or it simply may not have been adequately observed yet to disentangle.

Radio 21-cm measurements of neutral hydrogen gas radiation from IC 1613 have lead to the conclusion that the total mass of this galaxy is approximately 4×10^8 solar masses, with about 18 percent of this mass in the form of neutral hydrogen. Thus, it appears to be, relatively speaking, a galaxy that is fairly rich in gas, much more so than our local system or M 31, which contain only 2–3 percent of gas.

Dwarf Elliptical Neighbors

An important new discovery regarding nearby galaxies was made in 1937 when a galaxy of an apparently new type was found on plates obtained in South Africa at Harvard's Boyden Station. This strange object, as first described in the publication announcing its discovery, appeared to be unlike anything that had been seen before. Called the Sculptor system, after the constellation in which it was found, it was an extremely inconspicuous, though large, aggregate of very, very faint stars (Fig. 85). It appeared to have a roughly circular outline and a smooth distribution of stars, and on the discovery plates the stars all seemed to be of about the same magnitude, very near to the faint limit on the plate. At first it was not clear what it was, but subsequent study at Harvard and at Mount Wilson on plates taken with the 100-in telescope definitely proved it to be a distant galaxy of stars. The brightest stars were red, and

Fig. 85. Negative photograph of the Sculptor dwarf galaxy. (Cerro Tololo photograph, 60-inch telescope.)

this showed them to be similar in some respects to the globular-cluster stars. However, unlike a globular star cluster, it was extremely thinly populated with stars and intrinsically very large. Furthermore, it was not unique, for in 1938 a similar object was discovered in the constellation Fornax (Fig. 86) and in the 1950s four more were found in the northern skies.

Perhaps the most important first result in the study of these objects was the fact that they are, for galaxies, so extremely underluminous. Recent photoelectric measurements of their total brightnesses give them absolute magnitudes that are smaller than for any galaxies known previous to their discovery. That for the Sculptor system is only −10.9, so that the entire galaxy is fainter than the single brightest star in the Magellanic Clouds. These remarkably low total luminosities were a considerable surprise to astronomers in the late 1930s, who had assumed that galaxies were all of about the same total luminosity with only a few that are superluminous and a few that are subluminous. However, when the Sculptor and Fornax systems were added to the other known dwarf galaxies near

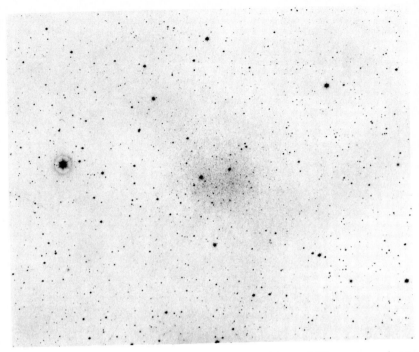

Fig. 86. Photograph of the Fornax dwarf galaxy. (Harvard Observatory photograph, ADH telescope.)

the Milky Way (NGC 6822 and IC 1613) it became clear that the distribution of luminosities of galaxies was decidedly not symmetric. Instead, there seemed to be an overabundance of the subluminous galaxies. Are most galaxies in the universe subluminous? This suggestion received further support from the discovery in the 1950s, when the Palomar sky survey photographs were first being taken, of the four additional dwarf elliptical galaxies, called Leo I, Leo II, Draco, and Ursa Minor (Fig. 87). Table 2 lists these six strange,

TABLE 2. *Dwarf elliptical galaxies.*

Name	Diameter (light years)	Distance (light years)	Absolute visual magnitude
Sculptor	3,900	275,000	−10.9
Fornax	10,000	600,000	−12
Leo I	3,000	700,000	−11.4
Leo II	2,100	700,000	−9.8
Draco	1,700	220,000	?
Ursa Minor	3,900	220,000	?

inconspicuous galaxies and describes many of their properties.

Detailed studies of the brightest stars within them became possible when large telescopes could measure (both photoelectrically and photographically) the colors and magnitudes of stars sufficiently faint. These studies showed that in fact the color-magnitude diagrams of these dwarf galaxies were very unlike those for young objects such as open star clusters, and unlike those for the irregular

Fig. 87.   The Ursa Minor dwarf galaxy. (Lick photograph, 120-inch telescope.)

galaxies near the Milky Way, which abound in young stars. Instead, the color-magnitude diagrams are almost indistinguishable from those of the Milky Way's globular star clusters, which contain exclusively old stars. Elliptical galaxies also contain such stars and it thus appears to be the case that the Sculptor-type galaxies are fundamentally similar in their stellar population to the elliptical galaxies that make up such a large proportion of the bright distant galaxies that we see. Unlike the bright elliptical galaxies, however, these newly discovered objects are very underpopulous (Fig. 88), with total masses estimated to be in the range of only 100,000 to a few million times the mass of the sun. On the other hand, a normal elliptical galaxy is more likely to have a mass that is tens of billions of times the mass of the sun. In fact, the masses of some of the nearby dwarf elliptical galaxies are no more than the masses of many globular star clusters in our Milky Way. The difference between them and the globular star clusters is mainly a matter of how far spread out the stars are within them. The dwarf elliptical galaxies are very much looser and of much smaller density than are globular star clusters.

The Sculptor system, for example, is so large that its major axis

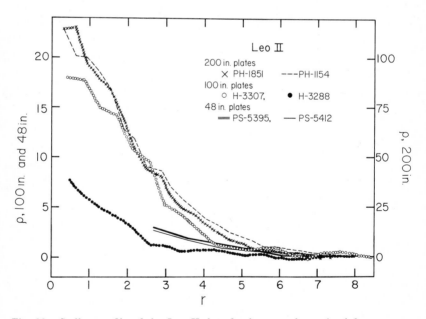

Fig. 88.   Stellar profile of the Leo II dwarf galaxy, as determined from counts of stars on plates from different telescopes. Distance from the center, r, is in minutes of arc. (From Hodge, *Astronomical Journal.*)

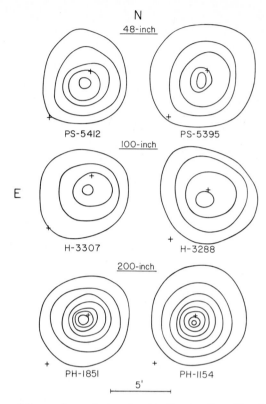

Fig. 89.   Sets of lines of equal star density for the Leo II galaxy as derived from six different photographic plates. (From Hodge, *Astronomical Journal*.)

has a maximum extent across the sky of more than 1.5°. This means that its true intrinsic size is about 7000 light-years, making it a very small galaxy though much too large for a globular star cluster. All of the local-group dwarf galaxies have been studied with regard to their structure and found to be perfectly elliptical, as nearly as star counts can show. Figure 89 shows plots of the density of stars in the Leo II system and illustrates its beautiful symmetry and the smoothness of star distribution in it. It contains no discernible dust or gas. Proof that these galaxies are essentially transparent and contain nothing but stars comes from the fact that all of them show much more distant galaxies shining through them, even at their centers.

Distances to these dwarf elliptical galaxies can be established by measurements of the cluster-type RR Lyrae variables that they contain. Five out of six have know RR Lyrae variable populations, the Sculptor system having so many that it is estimated that the total number of such variable stars must exceed 700. Only the Fornax galaxy, which is in the southern skies and a difficult object for northern telescopes to examine, has no known RR Lyrae variables. It is expected that the large telescopes now being built in the Southern Hemisphere will provide information on its variable-star population soon. For the rest, the RR Lyrae variables turn out to be completely normal in their properties. They behave so much like RR Lyrae variables in globular clusters of the Milky Way that we have very little doubt that they are intrinsically the same objects and therefore that we can obtain reliable measures of the distances of these objects simply from their apparent magnitudes. The distances range from a little over 200,000 light-years for the nearest (Ursa Minor and Draco) to about 800,000 light-years for the most distant (Leo I and Leo II).

The only member of this group that contains recognized clusters is the Fornax system, which is also the most massive and luminous. It contains six recognized clusters (Fig. 90), all of them fairly typical

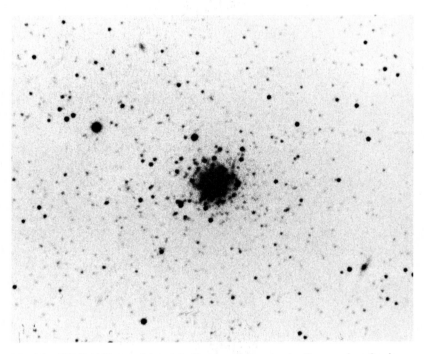

Fig. 90. NGC 1049, a bright globular star cluster in the Fornax dwarf galaxy. (Palomar photograph, 200-inch telescope.)

globular clusters, with stars that appear to be virtually the same in their properties as those in the galaxy itself. The colors of these star clusters are found to be normal and to be consistent with the idea that they are like the metal-poor globular star clusters of the Milky Way. They are systematically somewhat larger than globular clusters in the Milky Way, and this can be understood in terms of the weaker disruptive force of the galaxy itself on these clusters as compared with the very strong disruptive force of the Milky Way, which allows only globular clusters with a very strong concentration of mass toward their centers to withstand the gravitational effects of the main galaxy.

Possible Members of the Local Family

In addition to the galaxies that are clearly recognized members of the local group of galaxies, there are several that may or may not belong to our family. In all cases the uncertainty is due to the fact that these are objects that appear to be very highly obscured by the interstellar material in our local Milky Way. Objects that have low galactic latitudes, and hence are close to the Milky Way circle, are obscured so that we have a great deal of difficulty in establishing their distances by the usual photometric criteria. Presently we believe that one of these possibilities is most likely a member of the local group, though a rather distant one. This is the large spiral NGC 6946, which has recently been studied photometrically. Apparently the reddening and extinction of light caused by the intervening layers of dust in our Milky Way has dimmed the galaxy and its brightest stars to the point that we misjudge its distance by a very large factor. It appears in fact to be just barely within the usually established limits of the local cluster, with a distance that is roughly 3 million light-years. In 1971 another obscured system called Maffei I was identified as a possible giant elliptical galaxy member. Others of the possible members are less certain. For example, there are several strange irregular galaxies, such as IC 10, the Wolf-Lundmark-Mellotte galaxy, and irregular systems in Leo, Sextans, and Virgo, all of which are subluminous and poor in stars and not yet sufficiently studied to establish precise distances (Fig. 91). From what evidence we have, it appears that most of these are somewhat beyond the usually established limits of the local-group boundary and are therefore probably merely

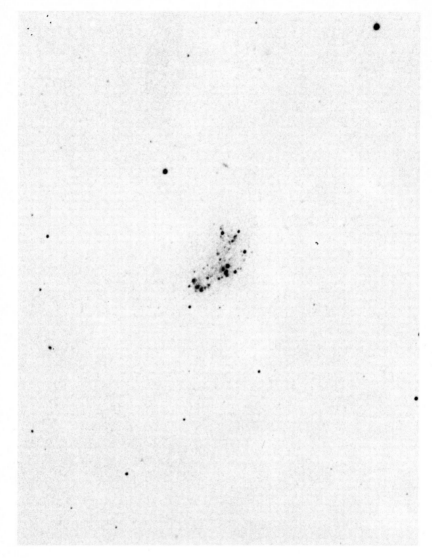

Fig. 91. A small irregular dwarf galaxy, GR 8, that may be a member of the local group of galaxies. (Lick photograph, 120-inch telescope.)

nearby small field galaxies unattached to our local clustering. There is of course a great deal of difficulty in establishing the complete population of our local group because there may well be galaxies that are entirely hidden by the Milky Way's obscuring matter. The only way that we might find of detecting such galaxies would be through their gravitational effect on the motions of the other galaxies of the local group. It is interesting that when calculations are made of the amount of energy (kinetic and potential energy) that our local group contains it is found that if the group is stable there must be more mass in it than we can account for by the galaxies that we see. Perhaps this mass exists in the form of one or more massive galaxies that we have not yet discovered because they are hidden behind the Milky Way. Perhaps, alternatively, this mass is in the form of intergalactic matter, either gas or dust or maybe even chunks of solid matter, that might pervade the space between the local galaxies. Or, finally, it may be that the local group is not stable and is collapsing or expanding so that the requirement that the energies balance is not necessary. We do not know the answer to this problem, but further study of the local-group galaxies and further searches for the effects of undiscovered members and intergalactic matter should throw light on this mystery in the near future.

6

Inside Galaxies

Weighing the Galaxies

In the few thousand years since an early scientist first discovered how to weigh things with a simple balance, mankind has extended his ability to weigh objects to an almost unbelievable range of masses. In physics laboratories scientists have found ways of measuring the masses of objects so small that they hardly exist at all, and in observatories astronomers have found ways of measuring masses of things as immense as whole galaxies of stars.

Spirals in Rotation

Observations by spectroscope show that the majority of galaxies rotate. The large spiral galaxies are truly whirling in space very much the way they appear to be at a casual glance.

The rotation is rapid on a cosmic scale, but exceedingly slow by human standards. Because of the immense sizes of galaxies, even though the stars in them may be moving with velocities as high as 200 miles a second (Fig. 92), it still takes about 100 million years for stars in a large spiral galaxy to make a complete circuit around its center. This is such a long time that there is no hope of detecting

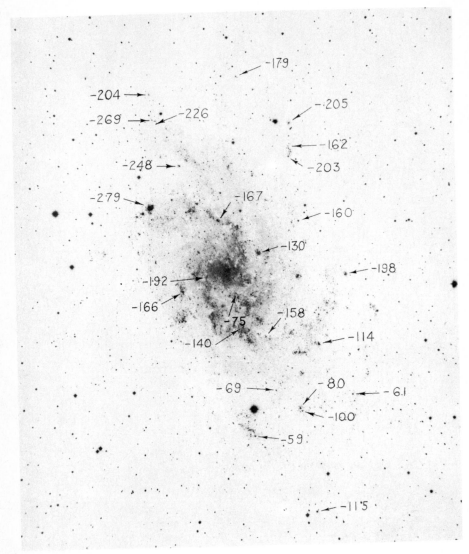

Fig. 92. Negative photograph of the spiral Messier 33, marked to show the positions and velocities in kilometers per second of the nebulous patches that were measured by Mayall and Aller in their study of the rotation of the spiral. (Lick photograph, Crossley telescope.)

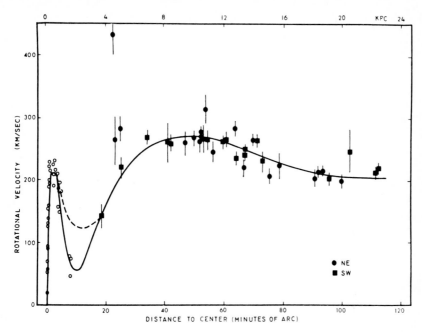

Fig. 93. Velocities of rotation for the Andromeda galaxy, Messier 31, determined from optical measures. (From Rubin and Ford, *Astrophysical Journal*, copyright University of Chicago Press.)

any motion of these stars by simply watching them during our lifetime. Instead, all of our information on the rotation of galaxies comes from the measured velocities of the parts of the galaxies observed by spectroscopic means. These velocities can be precisely measured because of the Doppler shift, which alters the wavelength of the light in a way that depends on how fast the stars are moving toward or away from us (Fig. 93).

The way in which a galaxy's stars rotate and the velocity with which they move depend to a large extent on the total mass of the galaxy. The gravitational control of the stars in a spiral differs, however, from that in our solar system, where essentially all the mass is concentrated in the central sun and the controlling force varies inversely as the square of the distance from the center. It also differs from what would prevail if the spiral were a huge discoid rotating as a solid with the force varying directly as the distance from the center. The actual situation is somewhere between these extremes and the composite law governing the motions of the stars in the galaxies changes with distance both from the nucleus and from the spiral arms. Therefore it is not possible to know just how gravity acts on a rotating spiral galaxy without first knowing how the stars themselves are distributed in space in the galaxy.

An Old Law on a Grand Scale

The first attempts to determine the total mass of galaxies used a simple law found nearly four centuries ago by Kepler. His laws of motion for the planets were interpreted by Newton to be manifestations of the universal law of gravitation. Newton showed that the relation found by Kepler between the period and the distance from the sun for a planetary orbit was a simple consequence of the gravitational pull of the total mass of the sun on the planet. Therefore astronomers interested in the total masses of galaxies attempted first to look at very distant portions of these galaxies, patches of stars so far away from the center of the galaxy that most of its mass was inside their orbits. This allowed them to apply the simple Keplerian law without their needing to know in detail the distribution of mass in the galaxy. When such a law is applied to a galaxy like the Andromeda Nebula or other large spirals, the masses of galaxies are found to be immense. For the Andromeda galaxy, for example, such a test shows the mass to be more than 100 billion times the mass of the sun. For M 33, the Triangulum galaxy, the mass is found in that way to be a few billion times the mass of the sun. And for our Galaxy it is found that the solar velocity, when tested with Kepler's law, indicates that its mass must be very similar to that of the Andromeda galaxy.

In retrospect it is perhaps not surprising to find that large spiral galaxies contain so much mass. If we make the assumption that the sun is a typical star in such a galaxy, then we could have predicted the total mass simply on the basis of the total measured luminosity from such galaxies. The Andromeda galaxy, for example, is bright enough to be seen without a telescope and yet is more than 2 million light-years away from us. This means that its total luminosity is more than 100 billion times the luminosity of the sun. From this it is clear that if the stars in the Andromeda galaxy are similar to the sun on the average we would predict that its total mass would also be a hundred billion times that of the sun.

Building a Galaxy to Fit

When observations of a large number of different positions in a galaxy give information on the velocities of rotation, it is then possible to obtain a detailed curve of velocity all the way from the central to the very outermost parts. Such a curve can be used to obtain a much more precise measurement of the total mass of the galaxy than can be got simply from Kepler's law (Fig. 94). Because the Keplerian formula applies only to a system with most of its mass concentrated at the center, a more complicated mathematics must be used for a galaxy that has much of its mass spread out. The modern method of deriving such results is to fit a model of a galaxy to the observational data. This involves constructing a variety of mathematical galaxies with a wide variety of distributions of mass. These then are used to derive the various rotation curves that would be observed if a given galaxy possessed such mass distributions. From the selection of different rotation curves one finds one eventually that fits an actual velocity curve for a galaxy.

Of course, an infinite number of different possible models might be derived that would fit a particular galaxy rotation curve, with

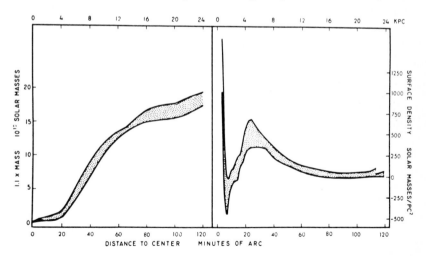

Fig. 94. Mass and density distributions for Messier 31, as deduced from measures of its rotation curve. The dotted areas indicate the range of possible values allowed by the observations and the mass models fitted to them. (After Rubin and Ford, *Astrophysical Journal,* copyright University of Chicago Press.)

its many uncertainties, and for that reason certain assumptions of basic principles are used to select among the various starting points that one might conceive. For example, mathematical forms of the mass distributions are chosen that are reasonable both from the point of view of the comparison with the light distributions in a galaxy and also from the point of view of the mathematics necessary to solve them. The easier methods are of course tried first. Perhaps the simplest kind of model that has been used is a "cookbook" model in which different ingredients are placed in a galaxy according to a set of directions that are based on other kinds of evidence, such as color measurements and photometry. For example, a very simple model of a spiral galaxy would contain a heavy spherical concentration of matter at the center to represent the nucleus plus a superimposed flat disk of smaller uniform density to represent the spiral arms and interarm material. This kind of model was used in the early days of model fitting, but it is not entirely satisfactory and more modern solutions of the problem involve much more complicated mixes.

Several dozen galaxies have been observed spectroscopically in sufficient detail to derive good rotation curves and for most of these it has been possible to find a model of mass distribution that can lead to an accurate total mass. On the average, galaxies have masses of approximately 100 billion times that of the sun, with the observed range for spiral galaxies extending from only about 20 billion all the way up to values of 400 billion solar masses.

Galaxies for which rotation is neither conspicuous nor observable include the elliptical galaxies and certain irregular objects. For these, the mass is more difficult to measure. In the case of elliptical galaxies, astronomers have found that it is possible to obtain a rather good idea of the mass from a simple measurement of the widths of the absorption lines in the spectra of these galaxies. The spectral lines are found to be much wider than the lines in an ordinary star and these widths are understandable as the result of the mixture of different velocities of the stars in the galaxy. By analyzing the line widths it is possible to discover what are the average velocities of individual stars in elliptical galaxies. From this it is possible to learn something about the gravitational attraction felt by these stars owing to the mass of the galaxy as a whole, and careful physical analysis of this effect can eventually lead to a

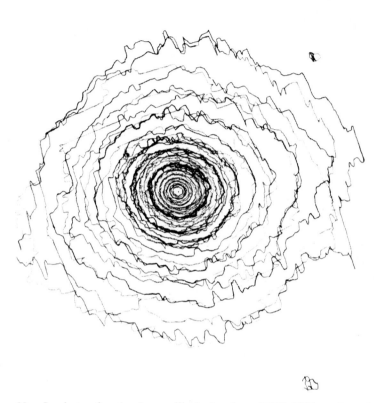

Fig. 95.   Isophotes for the large elliptical galaxy NGC 3379, a "standard" galaxy often used in photometry and spectroscopy.

determination of the total mass of the galaxy. Because of the difficulty of obtaining highly accurate spectra of faint objects, only a few elliptical galaxies have been weighed by this technique, but it is found to be an important source of information on these objects for which other methods do not work. For example, for M 32, the spiral elliptical companion of the Andromeda Nebula, it is found that the total mass is 4 billion times the mass of the sun. Thus, it is much less massive than most of the spiral galaxies. On the other hand, the large elliptical galaxy NGC 3379 (Fig. 95) has a total mass of 100 billion times the mass of the sun and it is therefore fully as massive as the large spiral galaxies.

Galaxy Twins

A further method of mass estimation involves the derivation of the relative motions in double galaxy systems. As in the case of the sun and the planets, the orbital motions of double galaxies provide a measure of their masses. From the work of various astronomers, including Erik Holmberg and Thornton Page, a great deal of information has been accumulated on pairs of galaxies from which masses can be extracted. Of course, not everything that one needs to know is known about galaxy pairs. We cannot wait long enough to determine the optical orbits of the galaxies around each other, because the periods are on the order of hundreds of millions of years. Therefore all we can do is measure their relative velocities with respect to our telescopes and their observed separation from each other. Holmberg and Page have shown that it is possible to eliminate uncertainties about the inclination of the orbit and its intrinsic size by measuring many pairs and then to obtain an average velocity for the masses of double galaxies. From a discussion of almost 100 pairs of galaxies, Page has shown that the mean mass for all of the system is 300 billion times the solar mass. When he separates out galaxies of different types, he finds the elliptical and S0 galaxies average more than this, 700 billion suns, while the spiral and irregular galaxies average only 40 billion suns. His computation of the ratio of mass to light (all in solar units) from these data shows that on the average the mass-to-light ratio is 40, but that it is only 3 for spiral and irregular galaxies and as large as 100 for the elliptical and S0 galaxies. The data on the mass-to-light ratio, incidentally, are almost as important to astronomy as the masses of galaxies because they tell us about the kinds of stars that exist in the galaxies of different types.

Invisible Signals

On the 25th of March 1951, Harvard scientists H. I. Ewen and Edward Purcell opened an entire new branch of astronomy. On that day they detected for the first time radio waves of neutral hydrogen from outside the solar system. This hydrogen radiation, with a wavelength of close to 21 cm, had been predicted as possibly observable 6 years before by H. C. van de Hulst in Holland, and its discovery at Harvard demonstrated not only that neutral hydro-

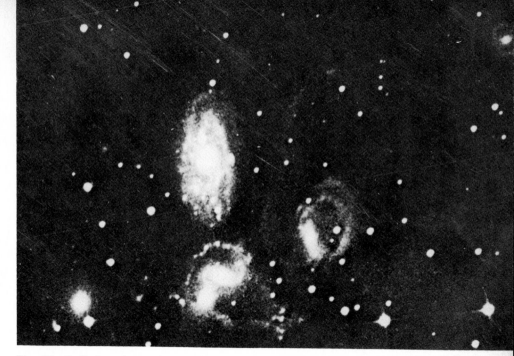

Fig. 96. Stefan's quintet of galaxies in Pegasus. One of the five galaxies is apparently not a true member of the group. (Mount Wilson photograph, 60-inch telescope.)

gen was detectable in our Galaxy but that the signals from it were so strong that it was clearly possible to map out the gas in our Galaxy and others more completely than could have been imagined. The radio line of neutral hydrogen therefore opened a new way of studying galaxies including our own, a way that penetrates through the screen of obscuring dust that has plagued astronomers who study our Galaxy with optical telescopes.

As an example of how greatly the study of the neutral hydrogen in galaxies has increased our knowledge we can look at the radio data that have accumulated in recent years for the Andromeda Nebula. Since the mid-1950s, when radio telescopes began to be large enough to look at individual galaxies in detail, a half dozen different surveys of the neutral-hydrogen content of the great nebula in Andromeda have been published (Fig. 97). These have all agreed in showing a considerable signal of hydrogen 21-cm radio radiation, invisible to the eye, and coming from the entire disk of the Andromeda galaxy. The signals come even from the faintest outermost portions of the galaxy, though they are concentrated toward the central regions where the brightest spiral arms exist. In fact, the neutral hydrogen that is emitting these signals seems to be arranged roughly in the form of a doughnut, with a hole in the center and

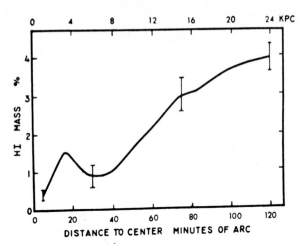

Fig. 97. The distribution of neutral hydrogen at different distances from the center of Messier 31.

with a maximum thickness (or more properly *density*) in a ring that has a radius of about 30,000 light-years. This gaseous ring is not a true doughnut, because the central areas show a minimum in the gas density rather than a complete hole and because the outer areas fade out so gradually that it is almost impossible to establish an outer bound. Furthermore, because the plane of the disk of the Andromeda galaxy is tilted steeply to the apparent plane of the sky, the "doughnut" is seen at a highly oblique angle.

By summing up over the entire signal of neutral hydrogen coming from the great Andromeda Nebula, it is possible to find out how much gas exists in this form. The total mass is computed to be about 4 billion times the mass of the sun. Although this sounds like a tremendous mass of gas, it amounts to only 1–2 percent of the total mass of the Andromeda Nebula itself. This percentage is actually a fairly typical value found for galaxies of type *Sb*, as has been discovered by examination of a dozen or so similar galaxies at greater distances. Galaxies of type *Sc* and irregular galaxies have a higher percentage of their mass in the form of gas (from 5 to 20 percent), whereas *Sa* and elliptical galaxies have very much smaller amounts of gas.

The neutral-hydrogen radiation at 21 cm is imporant for a reason other than its ability to tell us where the hydrogen gas lies. It is emitted in a narrow spectrum line of a known wavelength. Therefore, if the source of the radiation (the cloud of neutral hydrogen) is moving toward or away from us, we can detect this motion through the Doppler effect, which shifts the wavelength of the line toward either shorter or longer wavelengths. Because of this, therefore, it is possible to establish the motions of the neutral hydrogen gas in galaxies. For the Andromeda galaxy, the various studies of the 21-cm line have shown a clear rotation of the gas. The studies prove that the gas in the disk of this galaxy rotates in the same direction and at the same rate as the stars that can be observed with ordinary telescopes. Furthermore, because the neutral hydrogen gas extends out so far and can be detected at such large distances from the center of the galaxy, these studies have provided astronomers with an even clearer picture of the rotational properties of the Andromeda galaxy and have therefore better defined the curve of rotation used to determine the total mass of the system.

Unfortunately for purposes of model building of the Andromeda galaxy, the velocity measures of the neutral-hydrogen line show that it is a complex object. It has areas in which the velocities are quite different from what might be expected and which can be explained only by suggesting that the hydrogen gas not only rotates but also undergoes expansion, contraction, and localized distortion. An example of the "imperfection" of the rotation is the fact that on one side of the galaxy the rate of rotation is slightly different from that on the other side.

From the best fit of an idealized galaxy model to the radio observations it is possible, nevertheless, to establish with fair accuracy the total mass that is necessary to explain this rotation. The total mass is calculated to be 300 billion times the mass of the sun. This is a huge mass, even for a galaxy, and may even exceed the mass of our own giant stellar system. There are few galaxies known for which the mass is greater than that of the Andromeda galaxy, and therefore it is truly a giant among spirals. Such a huge mass is impossible to visualize; it we were to try to write it out in terms of familiar mass units, such as tons, the numbers are too big to mean anything. Who can imagine how heavy a galaxy really is if it weighs 600,000,000,000,000,000,000,000,000,000,000,000,000,000 tons?

The neutral-hydrogen radio radiation is not the only kind of invisible radiation that has made up our radio view of the universe. Galaxies also emit radiation with energy spread out over the entire radio spectrum. This continuum radiation does not allow us to study velocities because there are no discrete lines such as the 21-cm hydrogen line but it does give a kind of information that is difficult to get in other ways. The continuum radiation is from one of two sources. Most commonly it can come from a very hot gas cloud such as the bright cloud in Orion of our Galaxy. This is called "thermal" radiation and it not only gives us information on the temperatures and densities of these hot gas clouds but also allows us to detect them in areas that would otherwise be obscured from our view because of optically dense dust in front of them. The second kind of source is very much more unusual and caused a considerable stir when it was first identified in space. This is not a normal thermal type of source and the radiation given off is not emitted because of high temperatures of the atoms of the source. It is called "synchrotron" radiation, because it is commonly seen in high-energy physics laboratories that use nuclear-particle accelerators called synchrotrons. Synchrotron radiation comes about when particles (such as protons or electrons) are accelerated to extremely high velocities in a magnetic field. When charged particles such as these move in the magnetic field they are caused to spiral by the interaction between the electric field of the charged particle and the magnetic field in which it moves. If the particle is moving fast enough this spiraling leads to radiation. In the case of a synchrotron, the particles in the machine are moving with velocities very nearly equal to the velocity of light (but not quite, since relativity forbids exact equality with that velocity). Out in space the synchrotron radiation is emitted whenever an extremely energetic or explosive event occurs that can accelerate particles to similar extremely high velocities. Until synchrotron radiation was actually discovered in space, astronomers did not expect that any such event would occur, except possibly in occasional supernova explosions. However, as our knowledge of radio sources among galaxies has increased, it has become clear that synchrotron emission is fairly common and therefore that extremely explosive and energetic events occur with surprising frequency. This violence in the universe will not be discussed in this chapter, but because of its importance and the revolutionary changes it has brought about in astronomers' thinking about the evolution of galaxies a special chapter has been set aside for it (Chapter 7).

For ordinary normal galaxies the radio continuum radiation seems to be primarily emitted by common things. For example, in our own Galaxy most of this radiation comes from the flat disk and can be traced to individual hot gas clouds and supernova remnants.

The great Andromeda Nebula, being the nearest spiral, is a good example of a normal galaxy from which continuum radiation is observed. Several detailed studies of the continuum radiation have complemented the many studies of its neutral hydrogen radio character. For example, in 1968 the British astronomer G. G. Pooley used the Cambridge 1-mile radio telescope, a remarkable instrument giving very high resolution, to survey the area of the Andromeda galaxy for a study of its continuum radiation (Fig. 98). He found more than 200 different sources in the observed area. Only a few of these, however, turned out to be related to the galaxy itself, the remainder all being distant radio sources seen through the galaxy. The number of such distant sources could be established by surveys of the surrounding area, and was not unexpected because it is well known that the sky is extremely rich in distant discrete radio sources. At the longer wavelengths, Pooley found that almost all of the continuum radiation is nonthermal and it comes almost entirely from the nucleus and the brightest spiral arms. The nucleus is an especially bright object, suggesting the presence there of past violent events such as are discussed in Chapter 7. That the bright spiral arms are also conspicuous emitters of radio radiation is probably due to the fact that in these arms lie very large numbers of hot gas clouds, the thermal radiation of which contributes strongly to the emitted signals. The nonthermal radiation coming from the spiral arms, on the other hand, probably is emitted primarily by supernova remnants like the Crab Nebula of our Galaxy. It also may turn out to be associated with the presence of a stronger magnetic field in the spiral arms, a possibility that needs further exploration.

It is interesting to realize that the continuum radiation from the disk of the Andromeda galaxy is concentrated in the same ring or "doughnut" as the neutral-hydrogen radiation. This is a consequence of the fact that it is in this area that most of the star formation is going on. It is there that high temperatures and

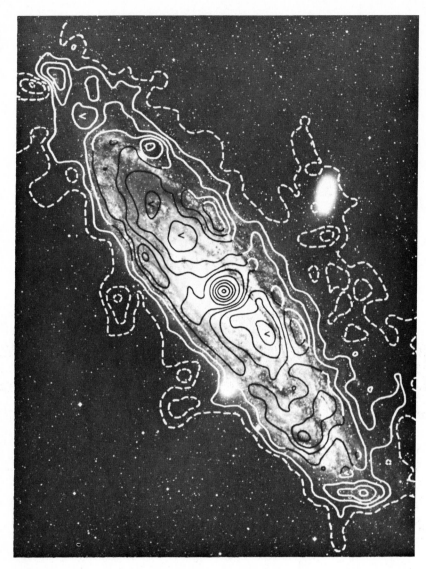

Fig. 98.   The distribution of continuum radio radiation from Messier 31 as measured by the Cambridge radio telescope. (From Pooley.)

supernova explosions result from the accelerated star-formation activity.

When galaxies farther away than that in Andromeda are observed, the high resolution that Pooley and others have obtained for it is not available. Therefore, our pictures of the radio views of other galaxies are very much more fuzzy. Nevertheless, the facts that we learned about the nearest galaxies such as the Andromeda Nebula are confirmed by what we can do with the more distant ones. For example, Fig. 99 shows the results of a radio study of

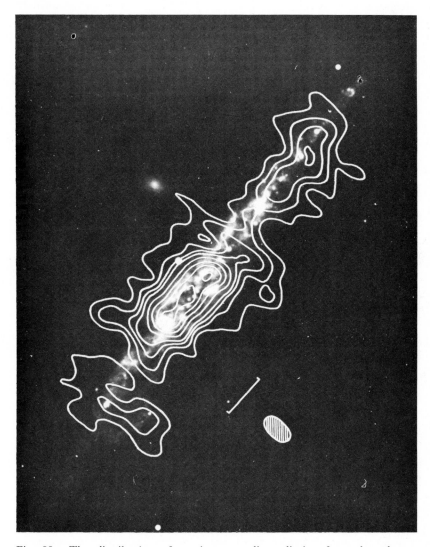

Fig. 99. The distribution of continuum radio radiation from the edge-on galaxy NGC 4631. The ellipse at the bottom of the figure shows the size of the beam "seen" by the radio telescope. (From Pooley, Mullard Radio Astronomy Observatory.)

the galaxy NGC 4631, examined in a way similar to that used by Pooley for the Andromeda galaxy. This galaxy, however, is seen almost edge-on and for that reason is interesting as a test case for the question of whether or not a "radio halo" exists. For this edge-on galaxy, with a visible thickness of only a few hundred light-years, any large, spherical halo of radio radiation extending significantly beyond the plane would be relatively easy to detect. Again using the Cambridge 1-mile telescope, Pooley observed this galaxy with many different separations between the antennas in order to give him the maximum possible resolution. The radio radiation from the galaxy was found to occupy a very narrow region centered on its optical image in the sky. The maximum thickness of the disk of the galaxy as detected in the continuum radio radiation is only 700 parsecs. This means that it must have a very flat distribution of objects emitting radio waves and that it has no large spherical halo such as sometimes seems to exist in galaxies.

The radio haloes of galaxies have been a matter of considerable discussion and dispute. A halo for our Galaxy was discovered in the early days of radio astronomy, but from more recent experiments it has been disputed. Its existence is very difficult to establish because of our internal viewpoint; we can't see the forest because we are in the middle of the trees. Because of the importance of the radio halo to problems having to do with the origin and the lifetimes of the cosmic rays in our Galaxy, it is very interesting to find out whether other galaxies, which we can see more clearly as a whole, have such haloes. It has been found by studies such as Pooley's of NGC 4631 that some galaxies have no detectable haloes whatever, but it is also found in the case of a few galaxies, such as that in Andromeda, that there is probably a radio halo. When enough different galaxies have been tested in this way, it is hoped that we will have a better understanding of why some galaxies have radio haloes, what implications the presence of a halo have for conditions in the galaxy, and what causes the halo in the first place.

Bright Clouds, Dark Clouds

Galaxies are not all stars. From their radio radiation we have found that a small but important percentage of most galaxies is made up of neutral hydrogen gas. An even smaller but also interesting portion of galaxies of most types can be detected optically with ordinary telescopes as bright gas clouds and dark dust clouds. For irregular and many *Sc* galaxies, it is often the case that the bright gas clouds are the brightest objects visible in them. And for other galaxies, especially spiral galaxies seen nearly edge-on, the dust clouds are the most conspicuous features in their photographs.

Bright gas clouds in galaxies (the H II regions) may hold the key to our understanding of the beginnings of stars (Figs. 100 and 101). They are bright because they are relatively dense masses of gas illuminated by extremely hot stars in their centers. Stars are forming there; so are many strange molecules, such as carbon monoxide, OH, and even formaldehyde (all detected by radio waves). There are pockets of gas that are very hot, while nearby are pockets of gas that are so cool that they can be detected only in the infrared. Cool, deep-red, dust-shrouded stars are found in these complicated gas clouds, and expanding groups of recently formed stars can be recognized in some. These many discoveries about the bright gas clouds in our Galaxy and others show quite plainly that it is in these special regions that most of the formation of stars occurs. It is here that nature gathers together the raw ingredients of the galaxy and puts the stars together, one by one, according to her still largely undiscovered recipe.

The bright gas clouds in our Galaxy are largely obscured from our view because of our location in the midst of its dusty disk. Only

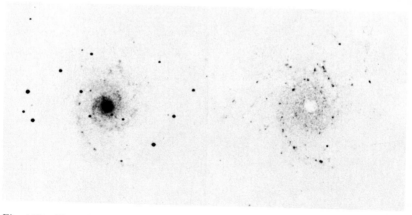

Fig. 100. Two photographs of the spiral galaxy NGC 628, on the left in yellow light, showing the distribution of stars, and on the right in hydrogen light, showing the distribution of hot gas clouds. (From Hodge, *Astrophysical Journal,* copyright University of Chicago Press.)

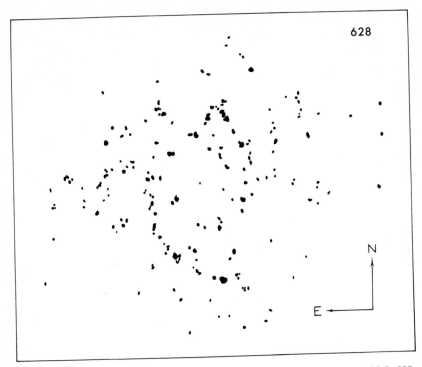

Fig. 101. A chart of the gas clouds discovered in the spiral galaxy NGC 628. (From Hodge, *Astrophysical Journal,* copyright University of Chicago Press.)

the nearest ones show up in optical telescopes; the more distant ones and most of the more interesting ones are visible only by radio. Therefore, if we wish to obtain an idea of where in a galaxy star formation is going on, it is better to look at the bright gas clouds in external galaxies. Most of the light given off by these clouds is in emission lines, emitted by the gas atoms because of the high temperature to which they are raised by the embedded hot stars. Therefore, astronomers search for them in other galaxies by limiting their telescopes and photographic plates to recording only the light from the region in the spectrum where there are such emission lines. For example, most studies of bright gas clouds in galaxies are made by recording the light of the emission line H$\propto$, a deep-red line of hydrogen gas. When an astronomer takes an H$\propto$ plate of a galaxy, he sees the galaxy's shape faintly spread out in its usual spiral or irregular form with bright spots scattered here and there over the object marking the location of the emission-line clouds.

Studies of the bright gas clouds in galaxies have shown that most of the star formation that we can detect in this way occurs in the spiral arms of the spiral galaxies. There are very few of these gas clouds between the arms and virtually none are found in the central areas of spirals. The highest density of star formation occurs for spiral galaxies about one-third of the distance outward from the center to the edge. This is something of a mystery because radio astronomers show through study of the same galaxies that the highest density of neutral hydrogen gas is two or three times farther out from the center. For some reason, therefore, it appears that stars are formed not in areas where the gas density is highest, but inside those areas, where stars and dust also lie thick.

Elliptical galaxies generally contain no bright gas clouds, though some have small bright nebulae in their nuclei. These are not normal H II regions, but rather are regions where high velocities prevail and where the atoms seem to be agitated more by the turbulence of the gas than by any imbedded bright stars. These strange nuclear clouds in elliptical galaxies are generally spherically distributed around the nucleus and often show spreads in velocity as large as 500 miles per second. Sometimes they are found to be rapidly rotating. They contain about 100,000 times as much mass as the sun, are a few hundred light-years in diameter, and have temperatures of about 10,000°K. Their most puzzling feature is the high degree of turbulent motion that they seem to maintain, because there is no clear way in which this turbulent motion can be kept going. Somehow energy must continuously be pumped into them.

In spiral and irregular galaxies the number of bright gas clouds is often very large. For example, for the Andromeda galaxy, Palomar Observatory astronomers found nearly 1000 H II regions. Hundreds have been found in the Magellanic Clouds and in M 33. Gas clouds are very much more numerous in irregular and *Sc* than in *Sb* and *Sa* galaxies.

Generally, spectrographic studies of these objects show that the temperatures are always on the order of 10,000°K and the density is always less than about 1000 electrons per cubic centimeter. The chemical composition detected by the study of the different emission lines in extragalactic gas clouds is almost always nearly the same as for that near the sun. In fact, it is generally found that the Orion Nebula, the brightest and most easily observed gas cloud near the sun, is fairly typical of gas clouds observed in our and other galaxies. Only in the very centers near the nuclei of galaxies are anomalies found that seem to be important. Here there are occasionally peculiarities in the chemistry and differences in the other physical properties.

The dark clouds in galaxies are due to accumulations of dust. Sometimes this dust shows up as long narrow absorption lanes cutting across the central regions (from our point of view) and sometimes as small patches of obscuration seen against the bright galaxy background. Elliptical galaxies (with a few exceptions) contain no detectable dust. Spiral and irregular galaxies, on the other hand, contain rather conspicuous amounts, especially in the central and intermediate areas. For some spiral galaxies, such as M 101, the dust seems to form lanes that continue the spiral structure right into the very center of the galaxy. The total amount of mass in the form of dust in the galaxies is small, only a very small percentage of the total mass of the galaxy. Nevertheless, it is important because the dust seems to be involved in an intricate and necessary way in the process of formation of stars.

Figure 102 shows the location of dust patches in the Large Magellanic Cloud. These are mostly faint and inconspicuous ob-

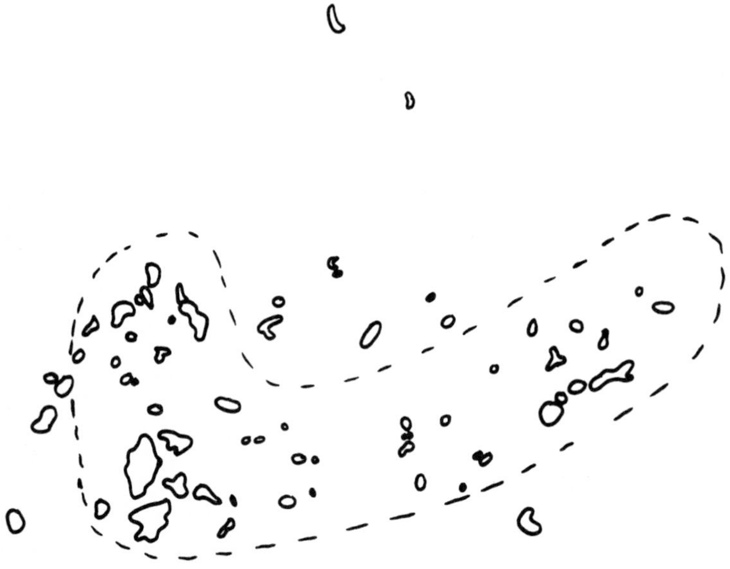

Fig. 102. The distribution of dust clouds (dark nebulae) discovered in the Large Magellanic Cloud. The dashed lines outline the main part of the bar of the Cloud. Compare with Fig. 18.

jects, just barely visible as depressions in the brightness of the Cloud. It is interesting to note that they occur primarily in areas that are also rich in bright gas clouds. For example, large numbers of small dust clouds are found near the giant 30 Doradus nebula and most of the remainder are near the bar of the Large Cloud. These dust clouds, which are fairly typical of such patches in galaxies, are 50 to 100 light-years across and the total amount of mass in the form of dust is found to average between 10 and 100 times the mass of the sun.

Among the many ways in which the properties of the dust in galaxies can be determined is the measurement of the polarization of the starlight from it. For example, for the Magellanic Clouds, it is found that the polarization has a remarkable orientation. Apparently the dust particles that cause this polarization are lined up in each Cloud in a direction that makes them all point approximately in the direction of the other Cloud, so that there seems to be a surprising parallelism of the dust grains in the two Clouds.

For spiral galaxies for which polarization measurements have been made it is found that the dust grains tend to line up along the spiral arms. This has often been used as a kind of evidence indicating that magnetic fields follow spiral arms and it is these magnetic fields that act upon the dust grains to line them all up in a particular direction. It is also important to note that the existence of polarization in the spiral arms implies the presence there of dust even in the outer parts, where the dust does not show up clearly on optical photographs. Dust is apparently found universally throughout spiral galaxies, though it is conspicuous only when it is gathered together in unusually dense patches and lanes.

Stars Young and Old

About a century ago the stars began to give up their many secrets to the curious scrutiny of man. First their distances, then their sizes, and more recently their composition have been unraveled through the ingenuity of astronomers of the past. But one of the stars' secrets seemed too illusive and obscure ever to unravel. How could man hope to discover the age of something so remote, so huge, so complicated, and so old as a star?

The answer came about 20 years ago, when theoretical studies of stellar interiors showed the way. It was found in the early 1950s that it was possible to explore reliably deep into the centers of stars entirely by using theoretical physics. Stars were found to be somewhat simpler in their interiors than on their surfaces and it was discovered that relatively simple and straight-forward physical laws applied to the conditions in their interiors. This discovery allowed astronomers to calculate the rate at which stars were using their fuel and from this to discover the life expectancies of stars. The combination of these theoretical studies with precise measures of the temperatures and luminosities of stars in clusters eventually led astronomers to the realization that the ages of stars could indeed be measured. They discovered the way in which sometimes small and sometimes great changes occurred in stars at different points in their lives. Stars at these various stages could be identified by their temperatures and luminosities. In this way many stars (and all star clusters) could have their ages measured.

The details of these calculations of star ages are fascinating and their story is told in *Atoms, Stars, and Nebulae,* another book in this series. For the purposes of this chapter we need only know that young stars are identified in star groups that are dominated by extremely luminous, blue stars of high temperature. Old stars, on the other hand, show up in areas where the stars seem all to be red, very faint, and cool. Therefore, in looking at star groups, one can determine immediately whether one is looking at an area of youth or an area of great age.

How Old Are the Galaxies?

Armed with the knowledge of how stars appear, young and old, one now can decipher the ages of galaxies. Ever since galaxies were first recognized as such 50 years ago, astronomers have speculated on their ages. These speculations were made primarily without help of very much physical information and therefore were often mere guess and conjecture. It was popular to suppose that galaxies had probably evolved from one Hubble class into another, and various interpretations of the Hubble sequence of galaxies were made in evolutionary terms. Perhaps, it was suggested, elliptical galaxies are young and as time passes they flatten out, take on a spiral shape, and eventually end as chaotic, irregular galaxies. On the other hand, it was alternatively suggested, the galaxies might have started originally in a state of chaos as irregular galaxies. Then as order began to take hold, they became *Sc* galaxies with spiral arms that gradually wound up until galaxies made their way through the Hubble sequence of spirals to *Sa* and finally to ellipticals, where order is complete. These speculations were important because they stimulated the imagination of the scientists who worked on galaxies to the point that they concentrated on means of checking the hypotheses of galactic evolution. Astronomers looked for every possible clue having to do with the ages of galaxies in order that one or the other of these speculations might be proved or disproved.

When finally it was possible to measure ages of stars, the day had come on which these fanciful ideas could at last be put to the test. Precise colors and luminosities of stars in galaxies of the different Hubble types were measured as soon as it was possible to do so. It required the largest telescopes in the world, combined with tremendous patience. At last in the 1960s, ages of stars and star groups in elliptical, spiral, and irregular galaxies had all been measured. The results did not go either of the ways that had been expected. Elliptical galaxies were not found to be the young ones, with the irregular galaxies ancient, and neither were the irregular galaxies found to be recently formed, with the elliptical galaxies old. Instead, it was found that all galaxies measured contained the same kind of very old stars. Apparently star formation began in all of them at a very distant epoch, measured to be approximately 10 billion years ago. The differences between galaxy types are not due to differences in total age of the galaxies but instead are differences in the way in which galaxies, since their birth, have carried on their production of stars. Elliptical galaxies contain only very old stars and therefore star formation in them must have ceased relatively soon after the birth of the galaxy. Spiral galaxies were more conservative about using up their stellar building blocks and they still contain some young stars and material (gas and dust) with which to form more stars in the future. Irregular galaxies have even more star formation going on at the present and an even smaller percentage of them is made up of ancient stars. So far, at least, all galaxies near enough to be studied in this way contain some of the same very old stars and therefore all galaxies that we can measure have the same age.

7

Violent Events

In the early days of radio astronomy, astronomers were at a loss to discover the identities of radio sources. The first radio telescopes were too small to be accurate enough to tell the exact locations of the sources they detected. Astronomers who examined the portion of the sky near radio sources using even the biggest optical telescopes in the world were unable at first to find objects that might be emitting the strong and strange radio waves. Except for the neutral-hydrogen sources, most of the small emitting regions detected by the radio astronomers did not seem to belong to any recognizable nearby object and therefore it was speculated that "radio" stars might exist that emitted only radio waves and little or no light at optical wavelengths.

However, as radio telescopes got bigger and better, more precise positions in the sky could be established for the brighter radio sources. Some of these turned out to be faint gas clouds such as the Crab Nebula and other supernova remnants in our Galaxy. But others were found to be much more remote and it became clear as time went on that the vast majority of the discrete sources of radio radiation detected must be outside our Galaxy, in the more distant reaches of the universe.

The first radio galaxy to be discovered optically was found after long and arduous searching on photographs taken for this purpose with the 200-inch telescope on Palomar Mountain. Radio astronomers called this source Cygnus A, because it was found to lie in the constellation Cygnus and because it was the brightest (that is, the strongest) radio source in that constellation. (It was in fact nearly as bright as the brightest of all radio sources in the sky, Cassiopeia A.) In 1954 Palomar astronomers Walter Baade and Rudolf Minkowski discovered an extremely faint object very near to the position that the radio astronomers gave for Cygnus A. Their attention was called to this object by its peculiar and mystifying shape. Its diffuseness suggested that it was a galaxy and not a star, but its shape was unlike that of any normal galaxy. It seemed to consist of two bright and connected condensations, both somewhat irregular. Surrounding them was a large, faint, elliptical halo of light. The distance to this strange object could be measured from its radial velocity, which was found to be about 10,000 miles per second. Such a high velocity for a galaxy indicates a distance of approximately 700 million light-years. Clearly this strange object must lie out among the distant galaxies.

In addition to its peculiar shape, Cygnus A showed another obvious anomalous feature. Optical spectra contained extremely bright emission lines, unlike those from normal galaxies. Much of the light from this radio source apparently comes from a large, turbulent, highly energetic gas cloud or clouds. The gas must be moving about with relative velocities as large as 300 miles per second. This, together with the apparent double nature of the galaxy, led to the concept of galaxy collisions.

Such a concept was very attractive at first, because it seemed to explain not only the strong radio emission from these objects but also the bright optical emission lines, since both kinds of energy would be radiated by the gas in two galaxies colliding with large velocity. Furthermore, other radio sources, as they became identified, also appeared to be colliding galaxies. A particularly attractive example was the source called Centaurus A, which was identified with the peculiar galaxy NGC 5128 (Fig. 105). On optical

Fig. 103. Messier 82, an exploding galaxy and radio source, showing the faint
outer filaments visible in hydrogen light. (Palomar photograph, 48-inch telescope.)

Fig. 104. The normal-looking radio galaxy NGC 1068, one of the galaxies discovered in the 1940's by Carl Seyfert to have an unusually bright and hot nucleus. (Palomar photograph, 48-inch telescope.)

Fig. 105. NGC 5128, a large and peculiar galaxy in Centaurus. Approximately 15 million light-years distant, with a velocity of recession of about 400 miles per second, this is one of the largest known radio galaxies, with a radio image nearly 100 times the size of its optical extent. (Harvard photograph, Rockefeller telescope.)

photographs this galaxy appears to be a combination of an elliptical galaxy and some late-type concentric spiral, and the thought at first was that this might be a collision of two galaxies which happen at this point to be completely superimposed on each other.

It was pointed out that galaxies in collision might pass through each other with only minor consequences. The stars in a galaxy are so thinly spread out and have so much space between them that few if any of the stars would collide if two galaxies should pass through each other. Only the gas in the galaxies would inter-act, heating up and probably dispersing as a result of the collision. It was thought that such collisions of galaxies might occasionally occur and that when the gas clouds in them collide the heating would be sufficient to cause intense radio and optical emission.

However, the collision hypothesis for radio galaxies was short-lived. As more and more radio galaxies were recognized and as more detailed studies of them were made, it was found that collisions simply could not explain them. First, it was calculated that the total amount of energy emitted by radio galaxies was far too great to be accounted for by a simple collision of two galaxies. Second, when the radial velocities of the two supposed colliding galaxies, such as Cygnus A, were measured they were found to be the same. And furthermore, most subsequently discovered radio galaxies showed no evidence of duplicity, but were rather single galaxies with usually some peculiarity in their structure.

A remarkable feature of Cygnus A, shared by Centaurus A and many other radio galaxies, is the very great difference in the radio size as compared to the optical size. The two optical condensations in its image are separated by only 2″ of arc. Radio measures at high resolution also show two separate condensations, but these are separated by over 100″ of arc. They are very nearly lined up along the same axis, but their scale is vastly different. Similarly, Centaurus A is a relatively small galaxy, about 15′ of arc in diameter. Its distance is difficult to measure but is estimated to be about 12 million light-years. This means that its apparent optical size is about 25,000 light-years. Its radio size, on the other hand, is vastly greater. Although there are two close radio sources only 12,000 light-years on either side of the center of the galaxy, the entire radio source itself is more than 100 times bigger than the optical galaxy. It extends over the sky for more than 10°, 20 times the size of the full moon. Its intrinsic size must be more than 3 billion light-years, indicating that it reaches out almost one-fourth the distance to us. The radio source is a huge object, many, many times bigger than any galaxy that has ever been detected optically.

Radio telescopes showed that the radiation coming from these remarkable radio galaxies was not normal thermal radiation but rather was synchrotron radiation. As described in Chapter 6, this kind of radiation is emitted by charged particles that move through magnetic fields with high velocities. The only conceivable way in which synchrotron radiation can be emitted from galaxies would be by some violent and explosive event that must have accelerated material from the galaxy to nearly the velocity of light and must have dispersed this material to very large distances. The total energy required to do this is staggering. Even now astronomers are hard put to find a mechanism of generating enough energy. In terms of ergs, the familiar small unit of energy used by physicists, a typical radio galaxy must be explained by an explosive event releasing some 10^{60} ergs. This is the equivalent of nearly 1 billion billion billion billion billion tons of TNT!

The Key at the Center

Clues to what is really happening in radio galaxies have come from two kinds of objects, one of them originally discovered long before radio galaxies and the other happened upon much more recently. In the 1940s the astronomer Carl Seyfert discovered the remarkable fact that a few galaxies (now called Seyfert galaxies) showed evidence of violent conditions in their nuclei. Seyfert galaxies had very bright, small nuclei, the spectra of which indicated the presence of gas at very high temperature. The gas clouds there apparently have high velocities of turbulence, up to thousands of miles per second, apparently resulting from some violent disturbance at the center of the galaxy.

For years the Seyfert galaxies were mainly regarded as curiosities that could not be explained easily, mere freaks among galaxies and not important to their understanding. However, when several of these few objects were found to be identical in position with strong radio sources, their relevance became clear. We now believe that there is a fundamental and important connection between the Seyfert galaxy phenomenon and the radio galaxies, either of degree or of evolution. Possibly Seyfert galaxies are galaxies experiencing a milder form of explosive disturbance than the more disrupted radio galaxies, where the radiation comes from the entire face of the galaxy or even larger areas. Or, possibly, the Seyfert galaxies represent the beginning of the explosive phase in a galaxy's history. It is estimated that approximately 1 percent of all galaxies are Seyfert galaxies and yet it is still not known whether this implies that 1 percent of all galaxies are "accident prone," galaxies that are always disturbed by explosive events in their nuclei, or whether it means that all galaxies experience these disturbances in their centers, but only about 1 percent of the time. Either of these possibilities, or some compromise possibility between the extremes, might be true.

Astronomers now have some clues to suggest that perhaps the latter possibility is most nearly the right one. For example, our own Galaxy, which is not one that would be an obvious Seyfert galaxy as seen from a distance, nevertheless has a moderately strong radio source identified with its center. Furthermore, there are peculiarities in the innermost portions of the neutral-hydrogen sheet of our Galaxy that suggest past disturbances in the center that have possibly expelled material from there. Other normal galaxies have

Fig. 106. The radio source known as 3C40, consisting of a radio galaxy in a moderately distant cluster. (Palomar photograph, 48-inch telescope.)

been identified also that have weak radio sources in their centers or similar evidence of disturbances there. The centers of galaxies perhaps provide the key to the history of the exploding radio galaxies.

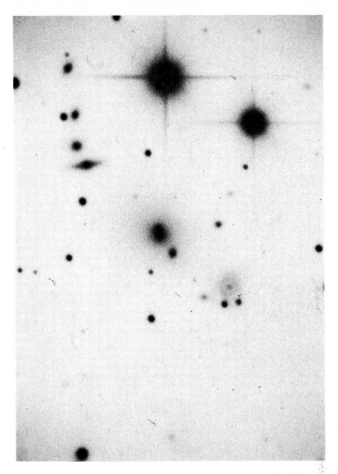

Fig. 107.   The distant radio galaxy known as Hydra A, a bright galaxy in a distant cluster. (Palomar photograph, 200-inch telescope.)

Quasars

The second event that provided further clues to the nature of radio galaxies occurred in 1960, when astronomers first discovered that many previously unidentified radio sources are associated with what appear to be unresolved stellar images. These were first called quasi-stellar radio sources, a name that has since been shortened to the term "quasars." They are remarkable objects. They seem to exist throughout space as far as we can detect and show completely unexpected and almost unbelievable properties. They are very small, estimated from their short periods of variability to be only a few light-years across or less, and extremely bright, the brightest having almost 100 times the luminosity of the brightest known galaxies. They emit tremendous amounts of both optical and radio radiation and show spectra that are somewhat like those of the nuclei of Seyfert galaxies. They appear almost to be dis-embodied nuclei of exploding galaxies, although none is near enough that astronomers can be sure as yet whether there is a galaxy surrounding it or not. (Any galactic disk might be obscured on astronomers' plates by the brilliance of the quasar itself.)

In the initial studies of quasars, there was a great deal of un-certainty about their location in space. Radial velocities obtained from spectra showed some of them to have extremely large red shifts, up to values more than five times as large as the largest red shifts detected for normal galaxies. However, no other method exists by which their distances can be estimated, and it remains therefore assumption that these red shifts are due to expansion as given by the Hubble law (Chapter 9). Some astronomers have suggested that these are not cosmological red shifts and that therefore the quasars are at different distances from what one would calculate from the Hubble expansion law. Perhaps they have been recently ejected from our Galaxy or from a nearby explosive galaxy such as Cen-taurus A. It has even been thought that quasars may have been ejected from distant disturbed galaxies that somehow produced them out of "new" material.

The reason that astronomers have taken such unorthodox views with regard to quasar distances is the very great difficulty that they have in explaining them if they are truly as large as they appear from the red shifts. Quasars are difficult to explain because of their extremely high luminosity, both at radio and at optical wave-lengths. This high luminosity implies a fantastic energy source, which has not yet been identified. The most promising explanations of quasars are based on the possibility of their being associated with what cosmologists call "singularities." Very massive objects that contract to very small size can, if they approach a certain limit given by general relativity, release large amounts of gravitational energy. The details of how this can be done are extremely compli-cated and are still being worked out. For example, among the attractive possibilities is a model in which the mass of the very massive object is contained in a rotating disk, the velocity of rota-tion of which is very nearly the velocity of light. Such an object could give up a great deal of energy, possibly enough to explain the quasars. There are many other models of quasars that may work. Included are models in which magnetic fields play an important part, models involving very dense clusters of stars or starlike objects, and models that hypothesize the existence in nature of "antimatter" which explosively reacts with matter.

It is too soon to say which of these lines of argument leads to the truth. Nevertheless, it is clear that there is probably an important link between the quasars and ordinary radio galaxies. The existence of peculiarly structured radio systems called "N galaxies" points to this link because they appear very similar to quasars, but with a faint disk of ordinary galactic stars surrounding the small nuclear source. There is, therefore, an empirical sequence that may in fact one day be shown to be evolutionary. The quasars, with their extremely small size and great brilliance, come first. Next come the N galaxies, with their radio and optical energy concentrated in an extremely bright nucleus, and following these are the Seyfert galaxies, which also have abnormally bright nuclei but which do not appear to be dominated in luminosity by their peculiar centers. Finally, the ordinary radio galaxies, with their radio energy spread out over their faces and beyond, may end the sequence.

Tempting as it is to picture this sequence in evolutionary terms, we have learned enough from our experience with ordinary galaxies and their evolution not to yield to this temptation carelessly. We may instead find that each of these peculiar types of radio sources is identified with a different kind of disturbance. Perhaps the differences between them are due to differences in their initial configuration, when they were forming as protogalaxies. We do not yet know, but it is very clear that such violence in galaxies is common and important. One of the great challenges in the immediate future in astronomy is the discovery of the source of this violence. Just what process in nature can produce enough energy to disrupt an entire galaxy?

8

Surveys of Deep Space

"Thickly populated district" would be an appropriate sign to warn extragalactic travelers in the neighborhood of our Galaxy. Within a distance of about 2 million light-years we have already found a score of others. Two of them appear to be a match for the biggest galaxies known anywhere in the whole extragalactic world, but mostly they are dwarfs. Once we have emerged from the region of our local group of galaxies, the population appears to thin out remarkably. If we extend the survey to 6 million light-years, increasing the volume surveyed by 27 times, we add only another dozen objects. Future careful searching will probably double that number because it will bring to light many dwarf galaxies, such as those that help make populous the local neighborhood. But we feel confident that the final roundup will still show that the average amount of matter per cubic light-year throughout the space occupied by the local group is at least ten times the average for the rest of surveyed space.

Do the more remote galaxies also belong to groups like our own? Many of them do, perhaps most of them, but some do not. The best way to seek an answer to this question is to examine the distribution of galaxies on the surface of the sky, and also, after estimating their distances, to examine their distribution in space. The survey plates by F. Zwicky and associates at Palomar and by C. D. Shane at the Lick Observatory reveal literally thousands of close groups of galaxies, many of them composed of only a dozen members or less.

We have already seen how difficult and uncertain is the measure of distance to a relatively nearby system like the Andromeda Nebula. The uncertainty of measurement will not decrease as we go farther from home. Nevertheless, with the use of the apparent magnitudes and angular diameters, we shall be able to get a preliminary idea of the distribution of galaxies in distance, as well as on the sky's surface. Then it can be judged whether the ordinary galaxy is isolationist or gregarious.

Census of the Inner Universe

The catalogue of clusters and nebulae compiled by the Herschels a century ago laid the foundation for J. L. E. Dreyer's *New General Catalogue* which, since its publication in 1888, has been the Holy Writ for astronomers working on nebulae, clusters, and external galaxies. The *New General Catalogue* (NGC) was followed in 1895 and 1910 with the first and second *Index Catalogues* (IC). Altogether the three publications include 13,226 entries. Several hundred numbers have been dropped in the course of later studies because they represented accidental double labeling, or pertained to double or multiple stars mistaken for nebulae. A few, wrongly listed as nebulae, were comets that have long since gone their way.

Many investigators have attempted to sort out the true nebulae from the external galaxies and distinguish the star clusters from everything else; then they make plots and studies of the distribution of the various types of objects catalogued in the NGC and the IC. A number of general conclusions have been correctly drawn from such plots, but the material has always been recognized as inhomogeneous. The Herschelian "sweeps" in the original search for nebulous objects were more complete in some parts of the sky than in others. Here and there the later surveys, using photographic methods, had dipped deep into space and brought up for the catalogues many faint objects in a small area. As a result, the undiscriminating plots of the entries in the NGC and IC occasionally seem to indicate a clustering of galaxies, when the true interpretation is merely depth in the census at that place, or unusual thoroughness.

Recognizing the unevenness of the NGC, the Harvard investigators of star clusters and galaxies undertook some 40 years ago (1930–1932) to employ a uniform series of photographic plates for a preliminary survey of all bright galaxies. For various practical reasons they decided first to make a new homogeneous listing of the galaxies brighter on the photographic scale than the thirteenth magnitude. The catalogue that resulted was published as part 2 of volume 88 of the *Annals of the Harvard College Observatory*.

For the study of nearby galaxies this Shapley-Ames catalogue has turned out to be very useful—as useful as it was laborious to prepare. The catalogue contains only 1249 galaxies, but 2 years were required for its formation, even though practically all of the necessary photographic plates were already in existence. The position of every object had to be checked. The photographic magnitude was measured on three plates. Many special sequences of standard stars had to be set up in order to make the magnitude estimates of similar quality over the whole sky. The angular diameters of the galaxies were measured, and something was done with the classifications.

All but 61 of the included objects were already listed half a century before in the NGC (but without useful magnitudes); of these 61 galaxies, 48 were in the IC, and 13 had not heretofore been catalogued. Although the thirteenth photographic magnitude was the desired limit, the catalogue is essentially complete only to magnitude 12.8.

The brightnesses of the galaxies were estimated on small-scale plates made with patrol cameras. On such plates it is possible to compare satisfactorily the images of galaxies directly with those of neighboring stars. With a few exceptions they look much alike. Reasonable care was of course exercised in the intercomparisons, and the results turned out to be somewhat better than the observers had expected. Since 1932, when Harvard *Annals 88*, No. 2, was completed, several investigations of some of the magnitudes have been made by more precise methods. Little change has been found necessary, except for the few galaxies brighter than magnitude 10, either in the zero point of the magnitude system or in its scale. If the makers of the catalogue had attempted to determine the magnitudes of the galaxies on plates made with larger telescopes, they would have gone astray because the stellar images and the nebular images would then have been too dissimilar for accurate estimating. The tiny patrol telescopes did something the giant telescopes could neither do easily, nor accomplish accurately, except through the use of elaborate accessories.

In Figs. 108 and 109 the distribution of the galaxies as shown by the "thirteenth-magnitude" census is illustrated in two different ways. The first figure is the reproduction of an Aitoff "equal-area" chart of the whole sky. Each plotted point represents the position of an external galaxy. The north pole of the heavens (Polaris region) is at the top of the diagram. The middle horizontal line is the celestial equator. The winding curve is the galactic circle, that is, the projection of the mid-line of the Milky Way against the background of the sky.

Two things stand out in this plotting (Fig. 108) of the objects of "88, No. 2"—the spottiness of the distribution and the almost complete absence of external galaxies from regions near the galactic circle. The discussion in Chapter 4 tells why galaxies are not seen

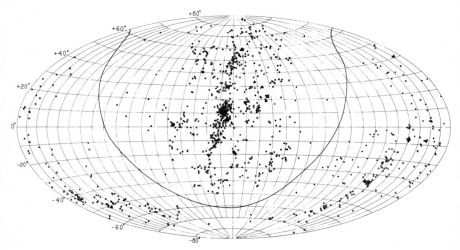

Fig. 108. An Aitoff equal-area chart on which are plotted, for the entire sky, the positions of the thousand brightest galaxies. (From the Shapley-Ames catalogue.)

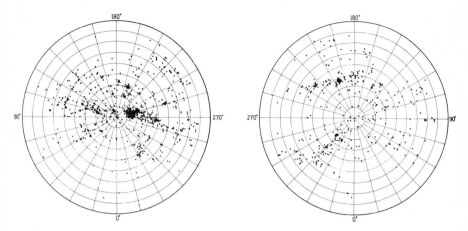

Fig. 109. Another chart of the brighter galaxies, also based on *Harvard College Observatory Annals 88*, No. 2, with separate diagrams for the north (*left*) and south galactic hemispheres. The irregularity in distribution of galaxies is the principal feature of this and the preceding figure.

near the circle, in lowest galactic latitudes. They are simply blocked out, or at least reduced to a magnitude fainter than 13, by the interstellar material near our own galactic plane. Effectively, only about one-half of the sky is clear. It is safe to assume that if the dust were absent there would be about twice as many galaxies brighter than the thirteenth magnitude as now appear in the catalogue.

In Fig. 109 the arrangement of these bright galaxies is shown in another kind of diagram—in galactic rather than equatorial coordinates—and the two galactic hemispheres are separated. The material is the same as for Fig. 108. This second aspect of the nearby galaxies again shows the spotted distribution, and the greater richness of the Northern Hemisphere. There are 823 galaxies on the north side of the Milky Way, and 426 on the south. The scarcity of the objects around the edges of the plots again emphasizes the effect of space absorption on the survey, for the outer parts of the diagrams correspond to low galactic latitudes.

The charts of distribution do not indicate any strong steady increase in number of galaxies with angular distance from the Milky Way. There is no obvious "concentration toward the galactic poles." If inherently there be such, it is smothered by the conspicuous irregularities in distribution. We shall consider the non-uniformities later, but first let us examine as best we can the distribution of these bright galaxies along the distance coordinate.

It is simple to compute how far the thirteenth-magnitude survey ($m < 12.8$) reaches into space. The distance d in light-years is given by a relation similar to that used in Chapter 2, namely, $\log d = 0.2$ ($m - \delta m - M$) + 1.5. If we set $\delta m = 0$, ignoring space absorption for the moment, and then take the absolute magnitude of an *average* galaxy as $M = -16.7$—a currently accepted value, when the objects are selected, as here, on the basis of their apparent brightness—we calculate that the distance of the faintest and most remote *average* galaxy in our catalogue is 25 million light-years.

If all the galaxies were of this average absolute luminosity, we could say that our survey reaches to the calculated distance. But they are not; and the reach is small for dwarfs, greater for giants. For example, dwarf galaxies with $M = -13.7$ would be, when the apparent magnitude is 12.8, at only one-fourth the distance calculated for the average galaxy; all such dwarfs, if between 6 and 25 million light-years distant, would be too faint to get into our thirteenth-magnitude catalogue. On the other hand, giant galaxies of absolute magnitude $- 18.2$ and apparent magnitude 12.8 would be twice as far away as the limiting distance for an average system. In other words, our catalogue contains several giant galaxies that are at a distance of some 50 million light-years, but it is quite incomplete for dwarf galaxies beyond 7 million light-years.

Making some allowances for space absorption, we shall say that *on the average* our bright-galaxy survey covers the first 20 million light-years, except for the low latitudes where we are blacked out by interstellar dust; in those dark regions the survey reaches anywhere from nowhere to nearly 20 million light-years.

While we are considering galaxy distances, it would be well to pause a moment for two incidental observations. The first one relates to units. We find that for the measurement of the universe it is as convenient to use the megaparsec for a unit of distance as the light-year. A megaparsec was defined in Chapter 1 as 1 million parsecs, or 3,260,000 light-years. It is the distance at which the radius of the earth's orbit (93 million miles) would subtend an angle of one millionth of a second of arc.

The other digression is to calculate the greatest distance we have now reached with the most powerful telescopes. For this purpose we may work near the galactic poles and assume therefore that space absorption is negligibly small; hence $\delta m = 0$. The faintest external galaxies yet recorded with the largest reflector (the Hale 200-inch telescope on Mount Palomar) are approximately of apparent magnitude $m = 23.0$, after an appropriate correction for red shift. Let us assume, reasonably, that among these faintest objects are some that are absolutely about as bright as the Andromeda Nebula, say $M = -20.0$. The assumption is very reasonable, but we cannot point to any particular image and say that that fuzzy speck records such a supergiant galaxy. We can only say that among a hundred specks at the margin of invisibility the probability is high that a few represent supergiant systems, of absolute magnitude -20.0.

The formula above yields the result:

$$\log d = 0.2 \ (23.0 + 20.0) + 1.5 = 10.1,$$
hence
$$d = 11{,}000{,}000{,}000 \text{ light-years} = 3{,}500 \text{ megaparsecs.}$$

We have therefore photographed galaxies in light that has been 110 million centuries crossing about 60,000,000,000,000,000,000,000 miles of space.

Since the publication of the Shapley-Ames catalogue of galaxies there have been several more recent attempts to gather together information on the brightest and more important galaxies in the sky. For example, de Vaucouleurs has published a revision of the Shapley-Ames catalogue called *The Reference Catalogue of Bright Galaxies* (University of Texas Press, Austin, 1964). He and Mrs. de Vaucouleurs collected all available information on luminosities and colors of galaxies in the Shapley-Ames catalogue and summarized all published data on their dimensions, their classifications, their radial velocities, and their positions. They also added several galaxies to the list and provided a comprehensive bibliography on the galaxies tabulated. This revised Shapley-Ames catalogue has come to be a frequently used reference for scientists who study galaxies.

Another important catalogue has also been completed in recent years. This is the catalogue of galaxies and of clusters of galaxies by Fritz Zwicky and several of his colleagues at the California Institute of Technology. Much more comprehensive in its scope than the Shapley-Ames catalogue, the Zwicky catalogue gives positions and magnitude data for over 30,000 galaxies and nearly 10,000 clusters of galaxies. It extends only a few degrees into the

Southern Hemisphere, but it is complete and comprehensive for the northern half of the sky.

Another survey of galaxies is that prepared by the Soviet astronomer Vorontsov-Velyaminov and his collaborators at the Sternberg State Astronomical Institute in Moscow. The Soviet catalogue has a somewhat brighter limiting magnitude than the Zwicky catalogue (about 15.1 versus about 15.7) but gives more information, including the dimensions, the relative intensities of the center and outer parts, the inclination of the galaxy to the plane of the sky, and detailed descriptions of the structure and morphological properties of each galaxy. The Vorontsov-Velyaminov catalogue includes nearly 35,000 objects.

Let us continue with the consideration of the distribution in distance of the brighter galaxies of the nearby universe. We know the position on the sky, that is, the right ascension and declination, of each one of these objects, with high accuracy; but because of the spread in the luminosities of galaxies, some bright, some faint, some average, we cannot easily locate accurately their positions along the line of sight. For a score or so of the nearer ones we can get the distances directly by measuring the brightnesses of their supergiant stars, when such stars are clearly distinguished. But for the hundreds of others all we can now do is to show, as in Fig. 110, for the material of the Shapley-Ames catalogue, the frequency of the total apparent magnitudes of the galaxies, and say what that frequency indicates about the distances of average galaxies. With space absorption neglected, which is reasonable for the higher galactic latitudes, we can compute that the average galaxy at the eleventh magnitude (the curves indicate about a dozen) would be at a distance of 3.5 megaparsecs; at the twelfth magnitude, 5.7 megaparsecs; at the thirteenth magnitude, 8.3 megaparsecs. (See the formula in Chapter 3; to get millions of light-years, multiply megaparsecs by 3.26.)

The smooth curve drawn in Fig. 110a indicates what the frequency of the apparent magnitudes would be if the galaxies were distributed with absolute uniformity throughout extragalactic space, that is, if there were no groupings, no systematic increase or decrease of number with distance, but always the same number of galaxies in a given cubic unit of space wherever located. This uniformity assumption, represented by the smooth curve, requires

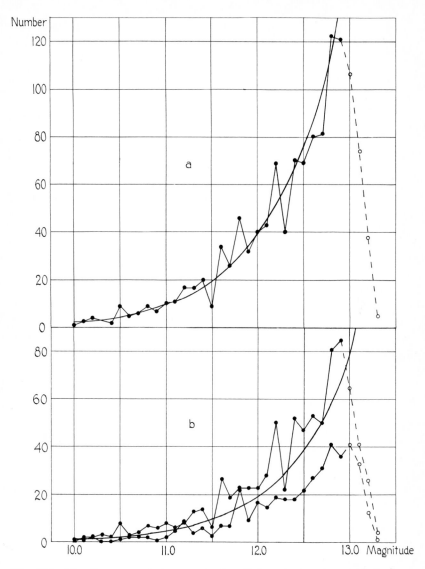

Fig. 110. The frequency of the magnitudes of the galaxies plotted in the two preceding figures. The vertical coordinates are numbers of galaxies for each tenth of a magnitude. The horizontal coordinates are the apparent photographic magnitudes of the galaxies. The upper figure refers to the whole sky; the lower figure, to the two galactic hemispheres separately, the southern being conspicuously the poorer.

that the number N of galaxies brighter than any given apparent magnitude m is related to that apparent magnitude by the formula

$$\log N = 0.6\,(m - m_1),$$

where m_1 is a constant called the space-density parameter.

To derive this simple but important formula, which we may call the uniform space-density relation, it is necessary to recall the formal definition of stellar magnitude m as 2.5 times the common logarithm of light intensity l. Numerically, the magnitude increases as the intensity decreases, so that $m \propto \log 1/l$. It is convenient to express the difference between two stellar magnitudes in terms of the ratio of the light intensities:

$$m - m_1 = 2.5 \log l_1/l,$$

and m_1 and l_1 may be taken as standards of magnitude and light intensity to which the other values are referred. Let us proceed to substitute numbers of galaxies for light intensities in this formula, and thus obtain a relation between apparent magnitude and the population of metagalactic space.

Since the intensity of the spreading light varies with the inverse square of the distance d from its source, we have $l \propto 1/d^2$, and

$$l_1/l = d^2/d_1^2.$$

The volume V of the space for which d is the radius (say the volume in space of a cone of diameter $1°$ and length d) varies, of course, with the cube of d, and therefore $d \propto V^{1/3}$ and

$$l_1/l = d^2/d_1^2 = V^{2/3}/V_1^{2/3}.$$

Therefore

$$2.5 \log V^{2/3}/V_1^{2/3} = m - m_1$$

or

$$\log V/V_1 = 0.6\,(m - m_1).$$

If space is uniformly populated with galaxies, their number N must increase with distance exactly as the volume of space increases with distance. Therefore $V/N = V_1/N_1$, where V_1 and N_1 may be taken as referring to the space defined by d_1 and l_1. Accordingly, we can write.

$$\log N/N_1 = 0.6\,(m - m_1).$$

If the standard limit m_1 of a "standard" survey of galaxies is so chosen that it corresponds to a distance d_1 and volume V_1 that are large enough to include just one galaxy, then $N_1 = 1$, and we have

$$\log N = 0.6 \ (m - m_1),$$

a formula that will be much used in our work of relating the number of galaxies to the apparent magnitude of the limit reached in various statistical surveys.

In practice it is found convenient to choose the unit volume as that covered by only 1 square degree of the sky. Therefore m_1 must be very faint in order to provide sufficient depth and volume to include on the average one average galaxy. The photographic magnitude 15.2, we shall see later, seems to be a good mean value of m_1, the space-density parameter, for the sky at large.

The space-density relation holds, by the way, even when there is a diversity among the actual luminosities of the galaxies, provided that the spread in the luminosities—that is, the relative numbers of dwarfs, normals, giants—is the same in all parts of the space considered. The formula is interpreted further in Chapter 10.

At first sight the deviation from uniformity seems not large. The smooth curve in Fig. 110*a* fits fairly well. But this is accidental, for in Fig. 110*b*, where the Northern and Southern Hemispheres are treated separately, the agreement is poor between observation and the uniformity curve. A clumpiness in the distribution of galaxies is suggested. Moreover, the surface distribution (Fig. 108 and 109) also emphasizes the grouping that prevails in some regions. Because of its significance in cosmology, we shall presently give further attention to the phenomenon of galaxy clustering, but first a look at the seemingly uninhabited celestial desert along the Milky Way—a desert for galaxies, even though it is the dominating metropolis for stars, nebulae, star clusters, and dust. This celestial Sahara is indeed so dusty that not only the galaxies but a majority of the remote Milky Way stars are dimmed out of our visual and photographic reach. The radio telescopes do a better job of penetrating dust clouds than either the human eye or the photographic plate.

The Region of Avoidance

The most pronounced unevenness in the distribution of galaxies is the rich population in high latitudes contrasted with the low population near the galactic circle. The region so conspicuously

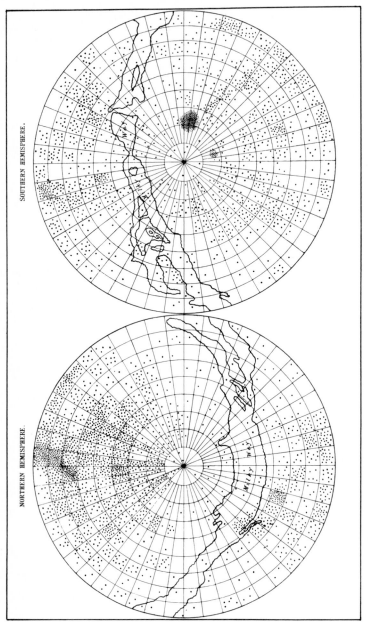

Fig. 111.   Proctor's early chart of external galaxies, which illustrates the important region of avoidance. The circular groups near the center of the right-hand figure represent nebulae and clusters in the two Magellanic Clouds. At the top of the left-hand figure is located the rich cloud of galaxies in Virgo.

"avoided" by the galaxies, known to workers in this field for a century, was pointed out most clearly by Richard A. Proctor over 90 years ago through his plotting of the objects recorded in Sir John Herschel's "General Catalogue." One of his illustrations is reproduced in Fig. 111.

The analogous, but much narrower, "region of avoidance" that pertains to globular star clusters of the galactic system was referred to in Chapter 4. It is now accepted that the narrow zone for clusters and the wide one for external galaxies arise from the same general cause—space absorption blocking out a part of the population. Region of obscuration would be a better name.

Shapley and Miss Jacqueline Sweeney made a special survey, on 400 long-exposure plates, of the distribution of 60,000 southern faint galaxies to show the extent of the light blocking in the southern Milky Way.

The region of obscuration for external galaxies is also shown by Hubble's sample-areas survey with the Mount Wilson reflectors (Fig. 112). His surveys, like those carried on at Harvard, not only reveal regions where the light of distant galaxies is completely blocked, but near the borders of the Milky Way they also help to measure quantitatively the amount of the absorption of light in space.

When we work in regions 30° or more from the Milky Way, we

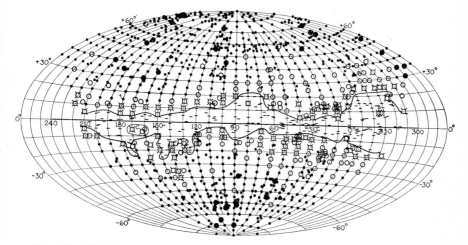

Fig. 112. Hubble's illustration of the region of avoidance, based on photographs taken with the Mount Wilson reflectors.

can, in the first approximation, ignore the space absorption. Certainly that is possible within 50° of the galactic poles, since the actual irregularities in the distribution of galaxies in the higher latitudes (Fig. 109) tend to conceal the evidence for whatever space absorption there may be in those regions.

The Virgo Cluster of Galaxies

The most conspicuous clustering shown in the thirteenth-magnitude survey is the one centered near right ascension $12^h\ 30^m$, declination $+ 12°$. This group lies chiefly in the constellation Virgo. There is also a considerable but looser grouping north of the Virgo cluster, extending about 40° through Coma, Lynx, and Ursa Major. In an analysis of the distribution of galaxies brighter than magnitude 12.7 (more than half of those in the Shapley-Ames catalogue), Katz and Mulders have shown that the chance is only one in 420 million that the arrangement of galaxies is random. In other words, the clustering is emphatically genuine. In the other hemisphere there is a bright group in Fornax, and there are others in Dorado and Grus. The Virgo organization merits a brief description since its relative nearness makes it particularly suitable for exploration.

Position and Population. It has been fortunate for astronomers, and for those who learn from them, that a supersystem of galaxies, much richer than our local group, is situated in a region favorable for detailed investigation. The Virgo cluster is near enough the celestial equator to be conveniently studied from all the important observatories, north and south. It is far enough from the galactic equator, with its troublesome space absorption, to simplify somewhat the photometry, as well as the measurement of distance. According to work by E. Holmberg, A. R. Sandage, and others, it appears to be about 11 megaparsecs away—a neighborly system as far as clusters of galaxies go, since its members are within range of both moderate-sized visual telescopes and small photographic cameras.

The diagram in Fig. 113 shows how the hundred brightest members of the Virgo group stand out when all galaxies brighter than the thirteenth magnitude are plotted for that part of the sky. The center of the concentration is near the middle of the triangle formed by the conspicuous stars Regulus, Spica, and Arcturus. No member

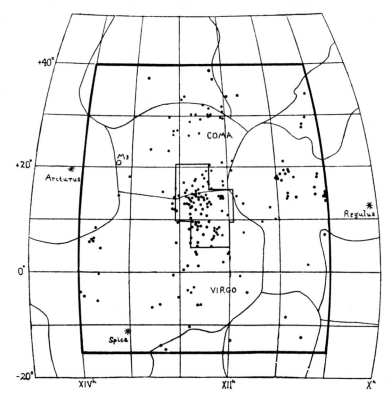

Fig. 113. The Virgo cloud of bright galaxies at the center of the Arcturus-Spica-Regulus triangle.

of the group is visible to our unaided eyes, because the more than 35 million light-years of intervening space has attenuated the light so that even the giant galaxies of the group can be seen only with the aid of a telescope.

When we extend our survey of the Virgo group to its fainter galaxies—to the fifteenth magnitude, for instance—we nearly double the population assignable to the cluster. It is difficult, however, to disentangle these fainter cluster members from the population of the general field.

Galaxy Types, Resolution, and Relative Sizes. Much attention has been paid to the Virgo group of galaxies, especially at the Mount Wilson, Palomar, and Harvard observatories. We know that about three-fourths of the bright members are spirals, and the others are mostly elliptical. There are but a few of the irregular Magellanic type among the brighter galaxies. An occasional spiral is somewhat freakish, but the majority belong to the category that we call *Sc*. At the Palomar Observatory a considerable number of these *Sc* spirals have been resolved; that is, individual supergiant stars within each galaxy, and clusters of giant stars, have been segregated, and their magnitudes estimated. Eventually all the *Sc* objects, from about magnitude 10.5 to 15, may be resolved, and also most of the *Sb* spirals.

The *Sa* spirals and the elliptical galaxies are more difficult to resolve, not so much because of the compactness of their structure as because supergiant stars are infrequent if not completely absent from them. Their ordinary giant stars, and of course the average stars like our sun, are too faint for the telescopic power of the present or near future. The brightest Magellanic-type systems would be easily resolvable if they were present in the Virgo cloud, for they, like our own Magellanic Clouds, would presumably be rich in supergiant stars. Among the dwarf members of the Virgo Cloud are many irregular galaxies that have been identified on Lick and Palomar photographs; they are too faint for the bright-galaxy catalogue.

It is of interest that throughout this Virgo assemblage, from the brightest to the fifteenth magnitude, the relative numbers of elliptical and spiral galaxies remain about the same. Also at any given brightness the over-all dimensions of the individual galaxies are found to be much alike for all galaxy types when the photographic exposures are sufficiently prolonged to reveal the faint outer portions of the elliptical galaxies.

Speeds and Masses. The radial velocities of about 100 of the Virgo galaxies have been measured, mostly by Humason at the Mount Wilson and Palomar Observatories. The group *as a whole* shows a recession from our Galaxy of the order of 700 miles per second; but there is much motion within the cluster, with a spread of more than 1600 miles per second in the velocities of the individual members.

Making use of the velocities, with certain assumptions concerning their meaning, Sinclair Smith has calculated what the total mass of the Virgo organization must be. It is enormous; and, when divided among the individual members now recorded, it indicates that they are each the equivalent of about 200,000 million suns. This seems like far too much mass for the amount of light produced, which averages but a few hundred million sun-power per galaxy. An alternative to accepting the great individual masses is to assume that much of the matter of the Virgo cluster is nonluminous dust in the spaces between galaxies. Or perhaps the speedy internal motions should not be attributed wholly to the gravitational inter-action of the individual galaxies, and should not therefore be taken as indicators of great mass. Further observation and further analysis are both important.

Spectral Types and Colors. The average spectral class of the Virgo galaxies, and of almost all others that have been sufficiently inves-tigated, is near that of the sun, *G*0. Some are of Class *F*, most are of Class *G*. Many of the irregular and *Sc* galaxies have spectral peculiarities that probably indicate the presence within them of very bright nebulae, or of groups of hot blue stars. They are the "blue" galaxies under investigation by G. Haro at Tonanzintla, and are importantly involved in W. W. Morgan's classification. It should be noted that a composite of all ordinary classes of stars, *O, B, A, F, G, K, M, N,* would be something like Class *G*.

The colors of galaxies have been measured by Stebbins and Whitford, Whipple, Seyfert, Haro, de Vaucouleurs, and others, with the uniform result that the color is about what one would expect it to be, considering the spectra, if there is no serious reddening of light in space. The colors, in fact, indicate high space trans-parency in the direction of the Virgo cluster, and perhaps they also indicate that there is not much space absorption within the Virgo cloud itself.

Are We in the Virgo Cloud? Returning to the diagrams showing the location of the Virgo cluster of galaxies (Figs. 108, 109, and 113), we notice that south of the main body of the cluster is an extension running nearly 30° toward the constellation Centaurus. Is this a part of the Virgo supersystem? If so, the over-all length is more than 8 million light-years. And to the northward are scat-tered bright galaxies, many of the same brightness and probably at about the same distance from us as the members of the cluster of galaxies in Virgo. Are they part of the same physical system? We should perhaps question, as Zwicky, de Vaucouleurs, and others

Fig. 114. A company of three tilted spirals in the far south, NGC 7582, 7590, 7599. (Harvard photograph, Bruce telescope.)

have done, whether we ourselves are not a part of this great cloud of galaxies that has a small condensation near us—the local family—as well as the much richer condensation in Virgo and the string of galaxies south and north. A local supergalaxy is suggested.

Evidence is growing that a large proportion of the galaxies within 40 million light-years are not free individuals in the universe, but rather are members of loose groups (Fig. 114). Are these sparse groups dissolving, or forming? We must wait and see. A billion years should suffice, or much less if our mathematical analyses of space, time, matter, and motion prosper.

The Fornax Group of Galaxies, and Others

Table 3 lists some of the information we have at hand concerning a score of bright objects in the constellation Fornax, which are so located with respect to one another that the law of chance is hard pressed if these objects are only accidentally near together. They appear to constitute a real colony of galaxies, mutually operating. In a number of groups such as this one we find that the brightest galaxy is of the elliptical type. Here it is the strongly concentrated NGC 1316. But there are also in the Fornax group dwarf ellipticals, similar to the unusual galaxy in Sculptor, which is one of the dwarf members of our own small family of galaxies. If the dwarf Sculptor galaxy were at the distance of the giant NGC 1316, and alongside it, the contrast would be most striking.

This contrast in luminosity emphasizes the fact that in our examination of the various groups of external galaxies, near or distant, we are always exposed to bias because we most easily study the giant galaxies. Our census of the population of a cluster of galaxies may be complete for the bright and sometimes for the intermediate objects, but in no group but our own do we yet know much about dwarf or subdwarf systems. We merely accept their probable abundance, and ignore them.

It turns out that from measures of apparent brightness only, and from the knowledge and techniques derived from studies of clusters of stars and clusters of galaxies, we can for the latter determine distances up to 200 million light-years and more, and know something of the error of the estimates. The error is not discouragingly large until we get out so far and so faint that we encounter the grave uncertainties (1) in the magnitude scales, (2) in the correction for the red shift, and (3) in the correction for space curvature, if any.

Thousands of clusters of galaxies are now known to be as rich

TABLE 3.   *Bright galaxies in Fornax.*

NGC number	Type	Magnitude	Right ascension	Declination
1316	Elliptical	10.1	$3^h\ 20^m7$	$-37°\ 25'$
1317	Spiral	12.2	3 20.8	-37 17
1326	Barred spiral	11.8	3 22.0	-36 39
1350	Barred spiral	11.8	3 29.1	-33 38
1351	Elliptical	12.8	3 28.6	-35 2
1365	Barred spiral	11.2	3 31.8	-36 18
1374	Elliptical	12.4	3 33.4	-35 24
1379	Elliptical	12.3	3 34.2	-35 37
1380	Spiral	11.4	3 34.6	-35 0
1381	Spiral	12.6	3 34.7	-35 28
1386	Barred spiral	12.4	3 35.0	-36 10
1387	Elliptical	12.1	3 35.1	-35 41
1389	Elliptical	12.8	3 35.3	-35 55
1399	Elliptical	10.9	3 36.6	-35 37
1404	Elliptical	11.5	3 37.0	-35 45
1427	Elliptical	12.4	3 40.4	-35 34
1437	Spiral	12.9	3 41.7	-36 1

as the nearby Virgo system, or richer. On plates made with the Palomar 48-inch Schmidt telescope G. O. Abell has indentified 2712 such clusters of galaxies. There are tens of thousands of groups that are as populous as the local group of galaxies, and also thousands of large distributional irregularities that strongly suggest immense physical associations. We observe, in fact, a basic tendency to cluster, whichever way we turn, and a high frequency of doubles. One is reminded of stellar analogies in our own Galaxy, where we find organizations of stars running from doubles through all degrees of grouping up to the myriad-starred globular clusters.

Faint-Galaxy Surveys at Harvard, Mount Wilson, Palomar, and Lick Observatories

The usefulness of the thirteenth-magnitude survey tempted the Harvard galaxy investigators to make a more far-reaching census. It seemed to be within reason to go to the eighteenth magnitude— not so deep that the galaxies, which come at the rate of one every cubic million light-years or so, would be practically innumerable, but deep enough that the returns should provide a large body of material—more than half a million galaxies (see Fig. 115)—for the examination of such cosmic problems as:

(1) The nature of the deviations from uniformity in the distribution of galaxies throughout the surrounding volume of space that has a radius of 200 or 300 million light-years;

(2) Statistics on the clustering of galaxies, and the bearing of such clusters on the development of the Inner Universe;

(3) The distribution of light-absorbing material in our own Galaxy, as indicated by the visibility of external galaxies along the Milky Way borders;

(4) The mean density of matter in explorable extragalactic space;

(5) The existence of significant large-scale gradients in the galaxy population of the space explored.

The survey covers the whole sky. The photographs for the Southern Hemisphere were made with the 24-inch Bruce refractor, then located on Harvard Kopje, near Bloemfontein, South Africa. The survey in the northern sky is based on plates made with the 16-inch Metcalf refractor, located at the Agassiz station on Oak Ridge, 25 miles northwest of Cambridge, Massachusetts. These

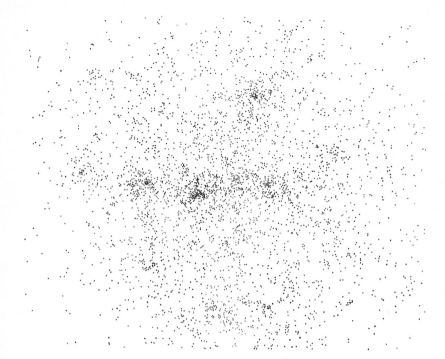

Fig. 115. This plot, made from a 3-hour-exposure photograph with the Bruce doublet at the Boyden Station, shows the distribution of 4000 galaxies heretofore unrecorded. The field center is at R.A. $13^h 25^m$, Dec. $-31°$.

instruments are of the same kind. They could be much improved with new-type lenses, but at the time of the census they were the two best galaxy-recording instruments (for survey work) in their respective hemispheres. Although they penetrated less deeply than the larger reflecting telescopes, they had the decided advantage in survey work of covering large fields. They each photographed satisfactorily something like 30 square degrees at a time, whereas the typical large reflector handles but a fraction of a single degree on one photograph. In 3-hour exposures on fast plates the Harvard instruments record stars somewhat fainter than the eighteenth magnitude—hence the name of the survey. But the galaxies are photographed satisfactorily for discovery only if they are as bright as stars that are approximately half a magnitude above the plate limit. The "eighteenth-magnitude survey," therefore, does not include all the galaxies to the eighteenth magnitude; magnitude 17.6 is approximately the limit for completeness.

More than half of the sky was examined on the long-exposure Harvard plates. Six hundred thousand new galaxies have been marked for the census, and magnitudes for tens of thousands have been measured.

Deeper than this eighteenth-magnitude survey were some of the sample-area countings by Hubble, who used the Mount Wilson reflectors. Long-exposure plates with the 200-inch telescope reached to the twenty-second magnitude. Hubble photographed areas so chosen as to give at least preliminary information concerning the nebular population throughout all the space within easy reach of his instruments. The two types of survey are complementary, and the pictures they give us of the universe are mutually consistent. They agree in showing that more than a billion galaxies are within the distance explorable by the greatest telescopes, but nearly half of these galaxies are concealed by low-latitude obscuration.

Two new surveys added importantly to our knowledge of the nearby universe. One is the work of C. D. Shane, using the 20-inch Carnegie astrograph at the Lick Observatory, and the other is the survey with the Schmidt 48-inch reflector on Mount Palomar. The former includes about twice as many galaxies as are recorded in Harvard's "eighteenth-magnitude" program, and the latter perhaps five times as many; neither covers the southernmost quarter of the sky.

9

The Expanding Universe

The mystery of the origin, destiny, and meaning of the physical universe inevitably incites to meditation all those who enter the spaces and times of the Cosmos. Whence did it come; whither it is going—and what is man that he writes books about it, and reads them? Certainly self-interest flavors his meditation, for, as the galaxies go, so go the stars, and the sun, and the sun's third planet earth with its superficial biology. But in this concluding chapter we shall evade the basic hows and whys, and continue to present the fragmentary observations and explanations that are slowly leading toward a finished picture of the sidereal world.

In entitling this chapter "The Expanding Universe," we have in mind, of course, the widely known observation that galaxies appear to recede from one another. If some day it should be convincingly shown that the red shift can be satisfactorily explained without recourse to the theory of a physical expansion, then the title above, we could say, refers to the unquestionable expansion of the universe of knowledge about the universe. Not only is that informational expansion unquestioned, it is amazing. The universe of galaxies is expanding at a rate that doubles the radius in a few thousand million years; but our knowledge of the universe doubles in one human generation. Our accelerated understanding encompasses not only galaxies and the anatomy of stars, but also the minutest particles, and their behavior in the microcosmos of molecules, atoms, and photons.

The machinery for research, in nearly all scientific fields, rapidly expands in variety and efficiency. New techniques evolve each year. Much inspiring accomplishment appears to be within our grasp. It will indeed be just as interesting to see how far human skill and understanding can go in this universe as to see what happens to receding galaxies, exploding supernovae, dissolving comets, and dying radiation.

The Space-Density Parameter

When we finished the preceding chapter we were 200 million light-years distant among the eighteenth-magnitude galaxies. It will be instructive to examine in some detail a sample of this distant realm.

The distribution of the faint galaxies over the central 9 square degrees of an average high-latitude Bruce plate, No. 20,309, of 3 hours' exposure, is shown in Fig. 116. The 659 small arrowheads point to the positions where an eyepiece examination of the original negative has shown new galaxies, heretofore unrecorded in any catalogue, probably never photographed before this plate was made. The large arrowheads locate the three external galaxies that had been recorded previously. They were, in fact, catalogued in the NGC and are conspicuous enough to be seen on this reproduction, although most of the fainter objects are lost in the process of reproducing.

The faint objects found on the plate are of various apparent magnitudes. In Fig. 117 is a diagram of the magnitude distribution. The vertical ordinates are numbers of galaxies in each small interval of brightness; the abscissas are the magnitudes. It is a conventional plot of the yield of galaxies as photographically we reach deeper into the universe. Magnitude 17.9 is the limit to which the galaxy count on this plate is complete.

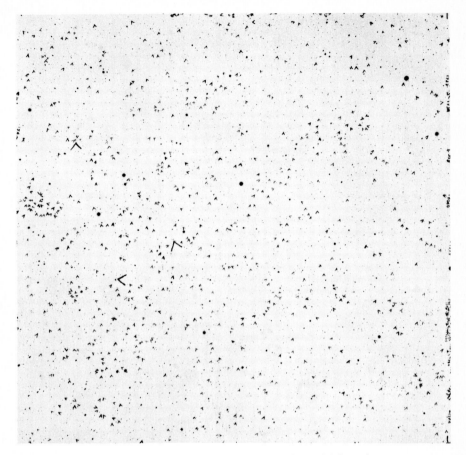

Fig. 116.   The central 9 square degrees of a Bruce plate of 3 hours' exposure, with large arrows indicating the 3 previously known galaxies and small arrows marking the 659 faint ones that now come into the census.

If space were uniformly populated with galaxies in the direction we have photographed on No. 20,309, the distribution would be as shown by the curved line. Obviously the fit of the actual count of galaxies to the uniformity hypothesis is not very good. There is an excess of galaxies around the eighteenth magnitude and a small deficiency around magnitude 16.5. The rise in the population graph is too steep for the uniformity hypothesis beyond magnitude 17.0. We may have reached into a metagalactic cloud of galaxies at a distance of about 100 megaparsecs. That would account for the excess, and the steepness. Another possibility, but not probability, is that the magnitude standards are at fault.

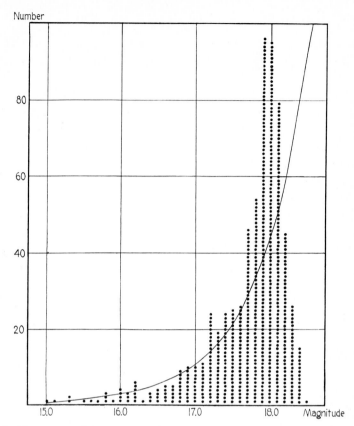

Fig. 117. Frequency of the apparent magnitudes of the newly recorded galaxies of Fig. 116.

To explain more technically in two paragraphs the practical procedure in this probing of space, let us plot as ordinate the logarithms of the numbers of galaxies brighter than a given magnitude, and not, as in Fig. 117, the numbers themselves. We get the diagram in Fig. 118. The straight line through the plotted points is the best representation we can obtain. It represents uniformity (the uniformity relation is derived and explained in the preceding chapter). The equation of the line may be written

$$\log N = b(m - m_1),$$

where N is the number of galaxies per square degree down to a given apparent magnitude m, and b and m_1 are constants defining, respectively, the slope of the line and its zero point. Ordinarily we call b the coefficient of the density gradient, and m_1 the space-density parameter. The latter is the magnitude down to which the survey must reach—here it is 13.9 for all 9 square degrees, or 15.5 for 1 square degree—in order to find on the average *one* galaxy per square degree. To check this definition, note that when $m = m_1$ the right-hand member of the equation becomes zero, and therefore N becomes unity. We determine m simply by counting galaxies of measured magnitude; but measuring the magnitudes accurately is not simple. If space is far from uniformly populated, the quantity m_1 has only local meaning and is not cosmically significant.

To illustrate the operation of the foregoing relation, let us suppose that there is at least approximate uniformity in the space distribution of galaxies, and therefore that $b = 0.6$. Then, if we find on the average 1 galaxy per square degree by going down to magnitude 15.5, as we do for Bruce plate No. 20,309, we should find on the average 4 when we get down to magnitude 16.5, 16 down to magnitude 17.5, and so on. Going in the other direction, it should require 4 square degrees to produce one galaxy of magnitude 14.5 or brighter, and 2048 square degrees (227 plates) with the average population shown in Fig. 116, to have one galaxy of the tenth

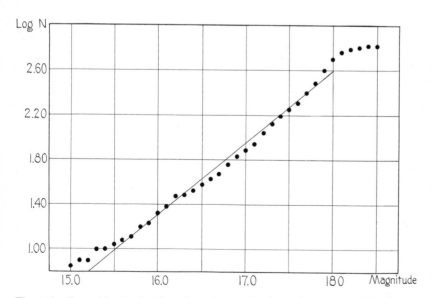

Fig. 118. Logarithmic plot, based on the results shown in the preceding figure. The slope of the line is important in the study of the population of space.

magnitude or brighter. Since 2048 square degrees is about 5 percent of the whole sky, we can compute from the faint-galaxy count on this one plate that there are all over the sky about 10 to 12 galaxies brighter than the tenth magnitude (allowing for dimming by interstellar dust)—and this number is not far wrong by actual bright-galaxy count, which gives 20.

For the Bruce plate represented by Fig. 116, the density-gradient coefficient is $b = 0.66$. If the coefficient b were exactly 0.6, the density of galaxies in space, in the direction covered by this photograph, would be exactly uniform; that is, every unit of volume of space would contain the same number of galaxies and matter would be uniformly distributed (in galaxy-sized chunks) throughout the space covered by the magnitude survey.

Since the density parameter m_1 really defines the number of average-sized galaxies in a unit volume of space, it is an important quantity, because its numerical value has much to do with the facts of cosmology—with the interpretation of the nature of space-time, the age of the expanding universe, and other questions of this sort. For example, its average numerical value is about 15.2; but if m_1 were 14.2, there would be four times as many galaxies in a given volume of space, the space-density of matter would be correspondingly four times greater, and the scattering of galaxies would be considerably less advanced than we now find it.

Anticipating the arguments of a later section, we note that the smaller the quantity m_1, the "younger" the expansion; the greater this parameter, the further along we are in our approach to zero density and infinite dissipation. Since galaxies are receding, and growing dimmer, m_1 increases with time. Some day, billions of years from now, m_1 will be fainter than magnitude 27, and it may then be difficult to photograph more than a score of galaxies, whereas now we can catalogue millions.

We are not quite ready to use the space-density parameter freely in cosmic interpretations because of suspected changes in its value from point to point in space, and also because of the uncertainty we have with respect to the masses of individual galactic systems. It is therefore still an uncertain leap from the number of galaxies per unit volume to the average density of matter in space. A fair estimate of average density is 10^{-30} gm/cm^3, or, in words, approximately a thousandth of a millionth of a billionth of a trillionth of the density of water.

Density Gradients

Is there any evidence of a center of the universe? Any evidence of an edge? Do our observations show any tendency toward systematic concentration, or systematic thinning out, in the number of galaxies, in the amount of matter, as we move in million-light-year strides across space?

We have found many cases of irregularities in the distribution of galaxies, but is there a general trend that would suggest a form or structure of the universe similar to that of a star cluster with dense nucleus and peripheral thinness, or analogous to the structure of our flattened-spiral Galaxy?

To save time, we go to the answer immediately, without bothering to present facts or arguments. The answer is "No bottom." There is no indication of a boundary; nor is there good evidence that there might be one if we went out far enough. If our measuring rods were longer than a billion light-years, or more sensitive to small density changes, we might find a falling off in some direction, or a clear trend toward some all-dominating nuclear cloud of galaxies; or we might glimpse clear evidence for a finite curved space; or, more likely, we might find as now no bottom and no excessive variation in the average frequency of galaxies.

The edge structure, if any, of the universe appears to escape us, but there is increasing knowledge of the internal anatomy. Already we have noted the giants and the dwarfs and the many types among the galaxies. We have photographed doubles, triples, multiples of many sorts. Thousands of groups are on record, and not a few very rich clusters of galaxies. In other words, we find not a rigorous identity among galaxies and not a dead uniformity in their distribution, but a prevalence of vestigial or embryonic organizations. We also find large clouds of galaxies of irregular outline—clouds that suggest chaos rather than orderliness. Certainly much time must flow before we attain, if ever, a smoothly populated universe.

When we find great unevenness in the distribution of galaxies in low latitudes, we may resort for the explanation to the hypothesis of light-scattering material in the interstellar spaces of our own Galaxy. But in many high-latitude regions we also find conspicuous irregularities in distribution that cannot well be attributed to clouds of intervening dust. We accept them as large-scale irregularities in metagalactic structure.

Figure 108 in Chapter 8 illustrates nonuniformity for the galaxies within 30 million light-years. As mentioned earlier, a number of other large-scale nonuniformities, larger than clusters of galaxies, have been found within the space now explorable. Long ago, for instance, it was noticed that the sky is much richer in bright galaxies in the northern galactic hemisphere (Virgo, Coma, Bootes, Ursa Major, Canes Venatici, and other constellations) than in the opposite southern galactic hemisphere (Pisces, Cetus, Sculptor, Aquarius, Pegasus, and others). Some investigators have taken the difference in the observed frequency of galaxies in the two hemispheres to be the result of the scattering and absorption of light within our galactic system. They argue that the number of recorded galaxies is greater on the north side because the sun is slightly to the north of the galactic plane, and therefore nearer the northern boundary of the supposedly uniform layer of scattering dust. There are several objections to this simple interpretation of the asymmetry in distribution. One is the lack of supporting evidence from the colors of stars and galaxies, although that objection could be met by "protective" subhypotheses. Another is the insufficiency of the absorption hypothesis, since the still greater differences in numbers of galaxies from one region to another in one and the same hemisphere remain unexplained. But the distributional inequalities for bright galaxies, and with them the absorption-inequality hypothesis, disappear in the Harvard and Mount Wilson faint-galaxy surveys, for at the eighteenth magnitude and fainter the galaxies appear to be equally numerous in the two hemispheres.

The observed difference between the north and south galactic hemispheres in the numbers of galaxies at the thirteenth magnitude (Fig. 109) is therefore only a structural detail of the nearby universe. The rich cluster of galaxies in Virgo contributes much to this inequality; but it persists even when we disallow the contribution from that prominent organization.

Some time ago Shapley undertook to examine quantitatively this north-south inequality, which is so conspicuous at the twelfth and thirteenth magnitudes, and apparently absent at the twentieth. The question of inequality for intermediate magnitudes and distances was examined. Galaxy counts were carried out in twelve high-latitude fields in the north and twelve in the south. Between the sixteenth and eighteenth magnitudes the values of the ratio of northern to southern galaxies (in half-magnitude intervals) were found to be as follows:

Magnitude	16.0	16.1	16.6	17.1	17.6
Ratio		1.25	1.11	1.44	1.55	1.09

Between the twelfth and thirteenth photographic magnitudes the ratio had been found to be about 1.4, even when the Virgo and Fornax clusters of bright galaxies were removed from the statistics. For all galaxies together, between the fourteenth and the seventeenth magnitudes, it was also 1.4. The northern galactic hemisphere appears to be 40 percent richer than the southern throughout this volume of space; farther out the difference disappears.

We should note that when we compare seventeenth-magnitude galaxies on one side of the Milky Way with the seventeenth-magnitude galaxies on the other, we are not dealing with short distances and localized irregularities. Such objects are in regions that are separated by nearly 500 million light-years. Whether the conspicuous population differences are to be accounted for by a great cloud of galaxies beyond the northern constellations, or are an indication of a major continuous south-to-north density increase, cannot now be determined. We first need to know the relative frequencies of the galaxies between the thirteenth and sixteenth magnitudes over large areas in the two hemispheres; and we shall also need to increase the number of regions examined for fainter galaxies before we can accept this south-north density gradient across the galactic plane as securely demonstrated.

Notwithstanding the lesser population of galaxies in the southern galactic hemisphere, which we find when intercomparing to the seventeenth magnitude the high-latitude regions of both hemispheres, the southern hemisphere has in the form of extensive clouds of galaxies at least two of the most conspicuous density irregularities. To one we have already referred in Chapter 5, where it was noted that even in relatively low galactic latitudes there is a rich background of faint and distant galaxies in the neighborhoods of the Andromeda Nebula and its associate, Messier 33. We cannot yet outline clearly the extent of this cloud of galaxies that extends away from the Milky Way through Andromeda into Triangulum, Pisces, and Pegasus. Tens of thousands of faint galaxies are involved. This string of galaxy clusters and clouds was first pointed out by

W. E. Bernheimer and his associates at the Lund Observatory; it and other possible superclusters are being further unraveled in Palomar galaxy surveys.

More clearly outlined than the cloud of galaxies just mentioned is the transverse stratum of galaxies that appears to be richest in the far-southern constellations of Pictor and Dorado, near the Large Magellanic Cloud. In this region, more than 200 million light-years away, it appears that the density of matter in space must be at least 50 percent higher than in other equally distant regions in the southern galactic hemisphere. Such differences are undoubtedly significant, since they are large-scale irregularities that must affect the large-scale operations of the universe; but as yet we cannot interpret their message.

In the course of a few years all the southern sky will be more thoroughly mapped, and diagrams of the distribution of several millions of galaxies will be available. Then we shall see if there are definite and smooth transverse gradients in density from one part of the sky to another. And when enough careful work has been done on the magnitudes of the galaxies, it should also be clear whether or not important *radial* density gradients exist—that is, notable increases or decreases in the number of galaxies as one proceeds outward in any direction. Already the magnitude work on individual plates has shown that there are local radial irregularities—both smooth and freakish deviations from uniformity in the population density as we travel outward, counting the galaxies as we go.

Notwithstanding the widely distributed clusters and clouds of galaxies, a large-scale population uniformity does seem to exist. For if we consider still greater volumes of space, lumping together all the material for half a sky at a time, we find an average uniformity; that is, the density-gradient coefficient is 0.6. This result is based on a study of more than 100,000 galaxies measured by the Harvard observers, and it substantiates the earlier results by Hubble on his sample-area studies at Mount Wilson. The outcome is important enough to merit repeating: when very large areas and depths of the sky are considered, the mean value of the density-gradient coefficient is almost exactly 0.6, and therefore the space density of galaxies appears to be uniform *on the average* within a sphere of perhaps a billion light-years diameter, notwithstanding the presence of the large and numerous clusters of galaxies, the enormous clouds of galaxies, and extensive areas of low population as shown by the Lick Observatory charts (Fig. 119).

If there is a general thinning-out of galaxy population with distance, or a thickening-up, it is so small, and becomes effective only at such great and dim distances, that we cannot be sure of it. The brightnesses of galaxies are hard to measure accurately when they are fainter than the eighteenth magnitude, where our standards of brightness are not nearly as safe as for brighter stars and galaxies. From surveys with the reflecting telescopes, Hubble found (after introducing corrections for red shift) what he believed may be a general radial gradient, which, over a distance of a quarter

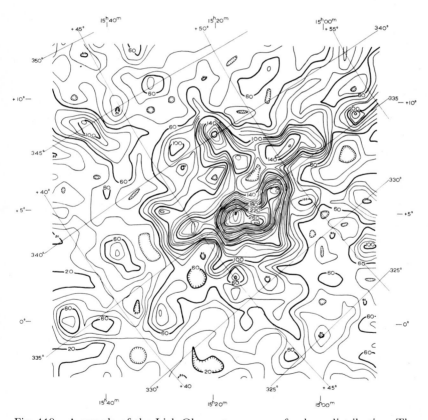

Fig. 119. A sample of the Lick Observatory maps of galaxy distribution. The great inequalities in the number of galaxies per square degree is illustrated; richest to poorest is about 10 to 1. The framing is shown in both galactic and equatorial coordinates. The center is at R.A. 5^h 20^m, Dec. $+5°$.

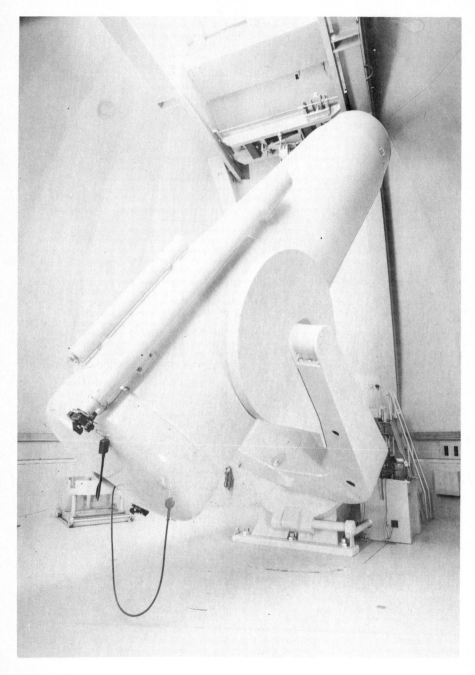

Fig. 120. The 48-inch Schmidt telescope on Mount Palomar, which has recorded some millions of faint and distant galaxies.

of a billion light-years, amounts to a density change of less than 20 percent. If fully established, such a radial gradient would be highly significant. Hubble naturally preferred to believe that the density is uniform. That would lead to simpler and more comfortable interpretations of the universe. Therefore, to eradicate the *apparent* increase in density with distance, and establish essential uniformity, he suggested the abandonment of the interpretation of the red shift as a velocity of recession, and the consequent abandonment of the hypothesis of the expanding universe; he would introduce in its place some new principle to account for the observed red shift.

But the evidence for Hubble's radial gradient has been thrown in doubt. The magnitude standards were systematically wrong at the faint end. Furthermore, if the Mount Wilson surveys, on which the deduction was based, are analyzed separately for the northern and southern galactic hemispheres, the density-of-population increase on the north is negligible. Moreover, the total amount of the increase with distance, as originally deduced, is not impressive compared with other more localized gradients, some of which certainly cannot be erased by revision of the magnitude standards.

Radial density gradients, of course, are more difficult to establish than transverse gradients. Shapley's work on about 75,000 faint galaxies in the southern galactic hemisphere showed a transverse gradient—a change in the frequency of galaxies per square degree in crossing a distance of some 500 million light-years—that is considerably greater than the general radial gradient originally suspected by Hubble, which led to his doubts concerning the existence of space curvature and of the alleged expansion of the universe. Later Hubble accepted the expansion as the proper interpretation of the red shift. Obviously we are not through with this business. The extensive galaxy surveys undertaken by Zwicky at Mount Palomar with the Schmidt telescopes (Fig. 120), and by Shane at the Lick Observatory with the Carnegie Astrograph (Fig. 121), will soon contribute importantly to our knowledge of population density.

The Motions of Galaxies

The foregoing discussion of density gradients in the cosmos ends with references, not fully explained, to the expanding universe and

Fig. 121. The Carnegie Astrograph of the Lick Observatory.

the relativistic cosmologies. It will be well to approach this subject by way of our knowledge of the motions of the galaxies.

Fifty years ago, when we were not sure whether the spirals were near at hand, among the fainter stars in our own Galaxy, or completely outside, it was natural that we should give them the cross-speed test for distance. The nearness of most nearby stars is easily discovered because their angular (cross) motions are large enough to be measured readily from year to year, or at least from decade to decade, or century to century. The more distant stars, just because they are distant, show little of this so-called proper motion or angular displacement, notwithstanding the fact that their cross speeds may be high. Proper motion, in fact, is a rough indicator of distance: small motion, far away; large motion, nearby (*or* excessive natural speed).

A. van Maanen with the large reflectors at Mount Wilson made valuable tests of the cross motions of the nearer galaxies as a part of his elaborate program on the proper motions and distances of galactic stars. His measures on the nuclei of spiral nebulae showed no appreciable cross motions for the intervals of time separating his earliest and latest photographic plates. If the plates had been separated by 1,000 or 10,000 years, the story would be different, because we are now pretty certain that the speeds of some of the galaxies are hundreds of miles per second, and in a long enough time measurable angular displacements must be possible.

If our present photographs of galaxies can be preserved for a few centuries, and duplicates then made for purposes of comparison, we should have valuable data on the cross currents in the inner universe, and we should have the means of analyzing the structure and dynamics of some of the nearer groups of galaxies. Moreover, we should be able, in a few thousand years, to learn much about the cluster of bright galaxies in Virgo, as we now learn about the bright clusters of stars in Taurus—the Pleiades and the Hyades.

The apparent fixity on the sky of the faint external galaxies from year to year is so dependable, because of their great distances, that we can reverse the usual procedure and, instead of trying to measure their motions with reference to our standard neighboring stars, use the faint galaxies as fixed points of reference in space against which to measure the proper motions of our stellar standards. A far-sighted program of this kind has been undertaken in recent years at the Lick Observatory with the Carnegie Astrograph, which was constructed for the purpose of making use of the galaxies.

Although crosswise motions of the galaxies are now immeasurable, the motions in the line of sight, derived spectroscopically through the well-known Doppler principle, have been measured for several hundred of the brightest systems, thanks to the powerful spectroscopes on the largest telescopes. The work is not simple; and the accuracy is, of course, not nearly as high as for similar measures on neighboring stars.

The radial velocities (line-of-sight motions) now available are largely due to the work of two specialists, Milton Humason of the Mount Wilson and Palomar observatories and N. U. Mayall of the Lick Observatory. One of the most outstanding contributions yet made with the California reflectors is Humason's and Mayall's measurements of the radial motions of very distant galaxies. Important pioneer work in this field was done by V. M. Slipher at the Lowell Observatory; and the McDonald Observatory astronomers have also contributed significantly, as has A. R. Sandage at Palomar.

Large reflecting telescopes are essential to the work because the light that arrives from extragalactic space is feeble. In order that radial velocities may be determined from the spectrograms, the feeble light must be spread out sufficiently by the prism to show recognizable features in the spectra.

Red Shifts and Cosmologies

The motion toward and from the observer is revealed in the faint spectra, according to the Doppler principle, by shifts of the spectral lines toward the blue and red ends of the spectrum, respectively. It was early discovered that except for a few nearby galaxies the spectrum shifts are all toward the red, and that the fainter the galaxy the more pronounced the red shift. Since faintness is associated with distance, it appeared, after sufficient observations had been accumulated, that the red shift was a fair indicator, if not an exact measure, of distance. Hubble derived the now well-known simple relation between the amount of red shift and the distance. Since we interpret that red shift for galaxies, as for the stars, as a direct result of motion away from the observer, the relation can be written as one between distance and speed in the line of sight.

Figure 122 diagrams an early version of the relation. It indicated

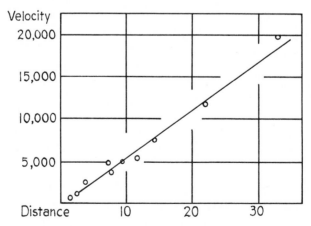

Fig. 122. The relation between the distances of galaxies (megaparsecs), and the red shifts of their spectra, the latter expressed as velocity of recession (miles per second). This figure is for clusters of galaxies relatively close to us, on a cosmic scale.

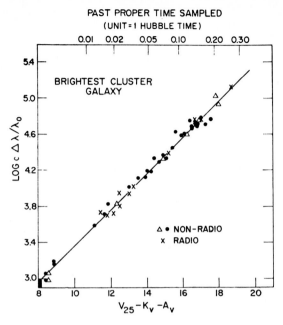

Fig. 123.   The diagram derived by Sandage to include the brightest galaxies in distant clusters, including some radio galaxies. Velocities are expressed in terms of red shifts multiplied by the velocity of light and the abscissa at the bottom gives corrected yellow magnitudes for the galaxies. At the top is the amount of time in terms of the Hubble age of the universe since the light that the telescope recorded left that galaxy.

that at a distance of 1 million light-years the galaxies recede at a speed of about 100 miles per second. At a distance of 2 million light-years, 200 miles per second; 10 million light-years, 1000 miles per second. But the revision of the zero point of the period-luminosity relation, described in Chapter 3, made revision necessary not only for distances of external galaxies, but also for the speed of expansion. The revising is not yet complete, since the scale of magnitude standards is not yet beyond correction. Tentatively we propose to accept a speed of expansion of 20 miles per second at a distance of 1 million light-years, and therefore a speed of 40 miles per second for a distance of 2 million light-years, and so on.

There was little tendency to question the interpretation of the red shift in terms of velocity so long as the measured speeds did not exceed a few hundred miles per second. Motions of that sort are known among the neighboring stars and are unquestioned. But when Humason's explorations reached objects more than 200 million light-years away, and the corresponding red shifts were indicating velocities of thousands of miles per second, some astronomers began to be uneasy; they wondered if out in those remote spaces something other than motion in the line of sight was producing the red shifts in the spectra. Velocities of 40,000 miles per second were attributed by Humason, on the basis of his spectrographic work with the Hale 200-inch reflector, to some of the faint galaxies, and W. A. Baum, with his photon-counter technique, went to higher speeds. Will those galaxies that we believe to be twice as far away as the remotest Humason measured move with speeds twice as great? And will those four to five times as far away show spectrum displacements to the red that would correspond to speeds of about 186,000 miles per second—the veloctiy of light?

The reason for the uneasiness is obvious. Scientists do not happily contemplate the possibility of real speeds greater than the velocity of light. But perhaps we are worried prematurely by this detail. The relation between distance and velocity, which has been shown to be approximately linear for the first 200 or 300 million light-years, is not yet accurately tested farther out. A deviation from the straight line seems inevitable; and already the 200-inch reflector on Palomar, equipped with the fastest possible spectrographic accessories, has shown that the red-shift relation, as we approach the billion-light-year distances, has become complicated. There is *tentative* evidence that the speed of expansion is now less than it was a billion years ago.

Not only do we need more of the extremely difficult measures of red shift for faint and distant galaxies, but we need, with equal urgency, accurate measures of magnitude for them. This second need emphasizes the importance of current research in light-recording techniques.

Before further comment on the limiting velocity is ventured, it would probably be best to await the accumulation of additional observations bearing on the distribution, brightnesses, colors, and motions of the galaxies in distant regions of the universe. But the impatient scientists have proceeded to look for other interpretations of the red shift. For example, could not the light of distant galaxies grow red with age? Those quanta of radiation that ultimately make our spectrograms have spent a million centuries or so traveling in space since they were emitted from the stars in the distant galaxies. The intervening space has some dust and gas in it, and everywhere it is being crisscrossed by the emitted radiations of millions of stars. In consequence, could not earthward-bound quanta of radiation lose some of their energy and thereby increase in wavelength and be recorded nearer the red end of the spectrum?

Or what of the hypothesis that long ago the atoms of all the elements were larger or smaller than they are now? The radiation from distant parts of the universe dates from a time in the remote past when the universe was younger. If in the interval of 100 million years or so the atoms have everywhere progressively changed in size or mass, there would likely be a corresponding change in the character of their radiations. We should not now without correction compare the shifted old light, which comes from atoms that were young when the radiation was emitted, with the new light from the old atoms of our terrestrial standards; the observed red shift may indicate not motion but the youth of the atoms whose antics produced the radiation. This, of course, is pure speculation, and not very intelligent.

There may be more sense to the speculative inquiry; Are we sure that the so-called fundamental "constants" of Nature (such as the velocity of light, the mass of the electron, its charge, the gravitational constant) are not in fact progressively variable over long intervals of time, such as are concerned in the measurement of the light of eighteenth-magnitude galaxies? If one persists in the conviction that galaxies, stars, planets, animals, and even matter evolve, why exclude categorically the primitive physical constants from the process of change, from developing with time and space?

It goes without saying that, if the constants are not constant, we need no longer strain outselves to interpret the red shift as due to a velocity of recession, or, for that matter, strive to interpret anything else on the cosmic scale.

Fortunately, there is one good reason for not now seeking furiously among these speculative alternatives in the hope of explaining away the red shift. Quite independently we have found in the theory of relativity an expectation that galaxies will scatter. The theory does not certainly predict the speed of recession, but an expansion of the universe is quite consistent with the general theory, which has been thoroughly tested in the nearer parts of the astronomical world and generally accepted. However much we may worry about the implications of the relativity theory at the bounds of measurable space, we are still pretty well satisfied of its validity and of its necessity near at home. The motions of Mercury's orbit, the red shift of light emitted from the surfaces of high-density stars, and the "bending" of rays of starlight measured at the time of solar eclipses—these are all well-known astronomical demonstrations that Einstein's slight modifications of Newton's gravitational principles are justified.

Although the effects introduced by the theory of relativity are quite trifling in the solar system, and except for a few problems completely negligible in the Galaxy, they become of rather major importance in the outer universe, and absolutely dominant when we try to figure out the total behavior of light, space, and time. Ambiguities arise at the boundaries, partly because of the lack of decisive observations and partly because the world now transcends our understanding—may always transcend it. We seek a satisfactory theoretical world model—something to visualize, if possible. To simplify the relevant and necessary mathematical and physical problems, certain assumptions and compromises must be made. We believe, for instance, that the truth about matter and motion in the universe lies between wholly motionless matter and wholly matterless motion; but how much of one, at this epoch, and how much of the other?

The necessary compromises create uncertainties and, supplemented by our present observational lacks, permit alternatives. We can logically deduce relativistic world models of various sorts. The universe on one model may alternately expand and contract. According to another, it may first have contracted as the stars were forming from a vague primordium, and now, with a reversal of trend, have gone into indefinite expansion toward the zero of density, the nothingness of heat. Or it may have erupted into an indefinitely expanding world from an infinitely old condition of stale equilibrium; or originated catastrophically from a single all-inclusive primeval atom some billions of years ago.

Since sufficient space cannot be taken here to present the relativistic cosmologies and the many interesting contributions and arguments concerning the bearing of relativity theory on various phases of cosmology, the interested reader is referred to semipopular expositions, starting with the early and very readable books by Eddington and Jeans (for example, Eddington's *Expanding Universe*, published in 1933) and including the following more recent books:

Bondi's *Cosmology* (1952), Schatzman's *Structure of the Universe* (1968), Hodge's *Concepts of the Universe* (1969), and Charon's *Cosmology* (1970).

Modern cosmologists have more powerful tools and much more information at their disposal than did the pioneers of cosmology, and yet they are still far from a consensus on the nature and origin of the universe. Obviously the picture is not yet clear. The finiteness of the universe is not established; nor is the contrary. Eternity may be nonsymmetric, differing in the forward aspect from the backward view. Time and labor will remove at least some of the ambiguities. In a pessimistic mood Sir James Jeans wrote: "As you will see by now, there is an absolute feast of hypotheses to choose between. You may pin your faith to any one you please, but you must not be certain about any. Personally, I feel very disinclined to pin my faith to any; it seems to me that it is still very open to question whether space is finite or infinite, whether it is curved or flat, whether the so-called constants of Nature change in value or stand still—if indeed any of these questions have any meaning." But the problems are not hopeless at all; and Jeans ended by quoting Robert Louis Stevenson that "to travel hopefully is a better thing than to arrive."

INDEX